Grow Your Own Food Organically

Grow Your Own Food Organically

BY

GARY NULL and Staff

THE HEALTH LIBRARY

ROBERT SPELLER & SONS, PUBLISHERS, INC.
New York, New York 10010

PRINTED IN THE UNITED STATES OF AMERICA

To

J.I. Rodale and Staff
Adelle Davis
Carlton Fredericks
Gayelord Hauser
Linda Clark
John Lust
Lelord Kordel
Carlson Wade
Herbert Shelton
Paul C. Bragg
Bob Hoffman
H. E. Kirschner
Linus Pauling
Ralph Nader

TABLE OF CONTENTS:

Grow Your Own Food Organically

INTRODUCTION

We are living in an age of ever-increasing awareness. More and more people are becoming seriously concerned with the quality of man's relationship to nature and beginning to question the fact that we have lost touch with the many gifts she offers. People are beginning to think organically, to explore the possibilities of returning to a more natural, less synthetic way of life. It's an exciting kind of revolution; based on the conviction that we can make our environment a healthier one and at the same time make our lives happier and heartier.

An abundance of information is now available in the fields of organic gardening, nutrition and ecology; but what about a really reliable compendium containing *all* these subjects? It is obvious that there is a direct, inextricable theme that binds these topics together and that they cannot be dealt with separately in a manner that is truly complete and satisfying. As things stand now, you would need perhaps twenty books to provide yourself with a good, overall view of the various aspects of organic gardening, nutrition and their bearing on ecology. Feeling strongly the need for one concise volume containing clear, helpful information, we have put it all together for you in one book, and this is it! What is more, this is a guide for *everyone*; the information we've included is applicable to city or country living and helpful to the 9-5 dabbler or the full-time farmer. We're eager to share with you our firm belief that organic gardening and the benefits it offers are not the exclusive province of those who live in a rural setting.

Anyone can grow his own food, and everyone should at least make an effort to do so. In fact, it is so easy that many of New York City's apartments should be fairly bursting with the fresh, wholesome foods an organic garden can produce—foods grown in your living room, prepared in your kitchen and served in your dining room! What could be better? Even a tiny apartment can boast a small, efficient garden if you follow the instructions we've outlined.

Our purpose in writing this book is two-fold. We are anxious

1

to spread the word that no one need be denied the pleasure of raising his own crop, no matter what his living situation; secondly,
we have attempted to give you a lucid idea of just exactly what
an *organic* garden is. So, with careful use of this book you are well
on your way to some exciting and healthful experiences in a realm
that you may not have realized is open to you!

Good luck and happy eating.

—Gary Null & Staff

WHAT IS ORGANIC GARDENING?

Modern urban man, so used to speeding along crowded highways, buying mass-manufactured products and wolfing TV dinners without a second thought, has lost contact with the sight, sound and feel of nature. There is a kind of painful irony in the fact that we have become so divorced from the very basis of our own existence. As much as we may try to ignore it, there is a yearning within us all to reunite ourselves with the flow of life from which we have sprung. We feel a deprivation and give it many different names; we seek substitutes at every turn, not knowing what it is we really are longing for. In answer to this urge we go to a park, a beach or a lush countryside, to experience the infinite wisdom and sanity of the ways of nature. To sense our oneness with this ultimate beauty is healthy, revitalizing and terribly necessary in a world that becomes increasingly synthetic and chaotic.

What happens when we haven't the time for a long country drive or leisurely seaside stroll? Is there a way to bring nature and all its accompanying beauty into our homes, even when that home happens to be at the busiest intersection of a booming metropolis? We say, certainly! It really is possible to surround yourself with the kind of greenery that is not only refreshing to your eye, but will provide you with some healthful and tasty eating as well!

The benefits of at-home gardening are countless, and many-leveled—even aside from the aesthetic considerations! Think for a moment what it would mean for you to become actively involved in the exquisite unfolding of the processes of nature—how soothing after a difficult day at the hectic city pace! We can't emphasize enough that the amount of space you have available is *not* evidence that gardening is impossible for you—even if it means your bathtub will be sprouting green peppers and you'll have watermelon vines twining around your stereo speakers. We've attempted to make clear in the following chapters that our instructions can be adjusted to apply to anyone's home set-up, whether it be small apartment, apartment roof, suburban home or a spacious country lot.

3

For those veteran green thumbs who already know the joy of growing their own, we hope to encourage a greater consciousness of the concepts behind the *organic* mode of farming if it has gone unconsidered. It is helpful (and exciting!) for anyone attempting to raise a garden to give some thought to its benefits as a valuable kind of "therapy"—guaranteed to raise your spirits right along with your cantaloupe! Read on.

Initiating a project involving the growing of vegetables will lead you to a new kind of familiarity with the natural processes we sometimes take for granted. You'll find yourself becoming aware of concepts like soil usage, conservation and the elements of building a fertile home for your garden. At a time when the need for ecology-minded people is a matter of do-or-die, this knowledge comes none too soon. So, gardening can be a learning experience, too—one you will want to share with your husband or wife or children—or anyone you care about who might become as gung-ho as you are once the project is set into motion. After awhile you'll find yourself wondering why everyone isn't anxious to tune in to the systems of nature as you have—and perhaps you'll have the joy of being the one to introduce them to it!

In our discussion of the merits of organic farming vs. the chemical method we have expounded further on the myriad benefits that await you as you embark on your new career as part-time organic farmer. If we keep reiterating all the pluses of organic gardening it's because we're so hooked on it ourselves!

WHAT IS ORGANIC GARDENING?

In essence, organic gardening is the raising of plants without any use of chemical materials, either as fertilizer or pesticides. The basic concept of organic gardening is to attempt to maintain a living fertility within the soil by constantly replenishing it with the minerals, plant and animal life that it needs so desperately. It has been proven that in order for plant life to thrive in soil, plant life *must be returned* to that soil year after year. Furthermore, experiments have shown that fertile soil itself is in large part composed of animal wastes and remains; the return of these elements to the soil is mandatory for healthy, living soil that will produce a plentiful harvest.

Early in the 19th century, Justus von Liebig, a German scientist, much in the spirit of the scientific and technological revolutions, began to experiment with the chemical relationships between plants. He came to the conclusion that specially developed fertilizers, designed to provide the soil with a powdered combination of phosphorus, nitrogen and potash would lead to a highly productive crop. What Mr. von Liebig forgot to realize was that soil is a living component existing not unto itself alone, but in the motif of a complexity of other living components. The single damning flaw of his theory was that he ignored the natural cycles of life—the unwritten laws beneath everything upon this earth of ours. Soil exists as one part of a quartet of natural components that must work in harmony to produce the substances required for life upon this planet. The scheme of life includes soil, water, animal life, and plant life. To disrupt the natural harmonies inherent in the workings of these elements is to bring disaster upon the entire framework of life forces. Von Liebig's plan to supply three chemical substances but ignore the vital organic portion of growing things could only pave the way for eventual disaster in the scope of natural resources. Man's presumptions to outwit nature, to hustle her, to take shortcuts, were bound to result in an ever-increasing deterioration in the bounty nature has to offer us.

When the word got out about von Liebig's experiments, the news was broadcast far and wide—a way to give nature a little push —higher profits! Commercial extravaganza! Work was begun to produce more powerful and potent chemical fertilizers. Voilà! A new and highly lucrative industry had been founded. The businessmen involved were so busy counting their profits that they didn't realize the farmers' loss of enthusiasm as the years passed. In a matter of two or three years, the initial increase of growth and yield had fallen off badly and plant life which continued to grow was noticeably weaker and less productive than ever before.

And so the organic movement, as we know it today, began as a reaction to the creation of chemical farming. Disillusioned, disgusted farmers left off purchasing the synthetic products and returned to a respect for the cycles of nature. Not all farmers of course—not by a long shot. The chemical fertilizer business boomed in spite of the obvious detrimental results ecologically speaking—the money was still rolling in.

In the way that von Liebig was the father of chemical farming, a Sir Albert Howard was the daddy of the growing campaign for sensible organic farming. Along with the other exponents of a natural method of raising crops, Howard predicted that the only-partial consideration of plant make-up would ultimately lead to a destruction of the soil's fertility by its very disregard for the earth's natural rhythm. He foresaw the tremendously adverse effects of chemicals and their malevolent contribution to widespread soil erosion and eventual loss of fertility—a process that would escalate as the use of synthetic fertilizers became more and more popular. Organic farmers visualized the accumulation of these poisons on our lands and projected the devastation that would follow as the years passed. The threat of a major upset in ecological balances brought on by a neglect of the soil's organic content spurred them on to pursue their studies and work in the field of natural farming. Their concern and persistence has served us well, in providing vast resources of information—but it is only recently that popular attention has turned to their work. The argument (waged most strongly by big business, no doubt) that organic farming was an obsolete, old-fashioned concept, obviously anti-progress, had scared countless farmers away from the possibility of experimenting with the findings of Sir Albert Howard and his associates. The idea that separating crops, soil, humans and livestock was unnatural and bound to be eventually harmful, did not stand up against the widespread publicity and hard sell of the chemical-farming industry.

To reiterate, the basic concept of organic gardening is that fertile soil can only be maintained by obeying the laws of natural cycles—that is, that all wastes, both plant and animal, be returned to the soil. Dust to dust, ashes to ashes! The workings of nature are indeed impressive in their success, in their eminent sanity and healthfulness. Isn't it about time man relearned a lesson from Mother Earth? Why not allow the natural cycles to work for us—to use our own gift of rational thinking to enhance these processes, instead of fighting a battle that we can only lose. It is not difficult—it only requires a little knowledge of natural systems. This knowledge can be exciting to you whether you have a three-acre plot or one lonesome mango tree in your living room. And, to comprehend the flow of nature is to grasp in great part an understanding of our own inner workings.

HEALTH AND ORGANIC GARDENING

It stands to reason (and it is woefully so!) that foods grown with chemical substances, withdrawn from their natural leanings toward the ways of a cyclical scheme of life, are bound to be unhealthy. Food that is grown on lifeless, chemicalized soil does not contain the nutrients and proteins which contribute to our health. "Eat your vegetables" becomes a kind of frightening mockery when one realizes that the very substances we are urging our growing children to eat are often soaked with harmful poisons and toxins.

With chemically grown foods we have a vision rather ghastly to behold—food (meant to *sustain* life) grown in a soil that itself is lifeless due to incomplete, artificial fertilization, and then (as if that weren't bad enough!) sprayed with DDT and similar poisons to deter garden pests. It is not that the mineral nutrients provided in chemical fertilizers do not provide some ingredients essential to plant growth that is detrimental; it is disregarding the fact that the organic content of the earth is its actual productive core that spells trouble. An abundance of plant material, at various stages of decomposition, insures that all materials—animate as well as inanimate will function harmoniously and efficiently. So, it is easy to understand why this partially nourished soil can only provide us with partially nourished fruits and vegetables which are, in their turn, only partially *nourishing*. To fail to realize the importance of the direct link between the health of the land and. the resulting health of the systems that consume the food is a good example of the kind of irrational thinking that has brought our environment to the sorry state it exists in now. Simple, clear thinking provides us with the immediate understanding that we are cheating our own bodies with these half-foods and we are poisoning ourselves with the very poison that chemical farmers coat their plants with. Again, there is a tragic irony in spraying foods to *protect* them when the ultimate result is that these foods endanger our own protection of bodily health. As determined voices of ecology become more and more insistent about being heard, we find it increasingly difficult to ignore the situation they so vehemently decry.

It is a Scientific fact that failure to return manure, crop residues, and other organic matter to the soil (the very basis of organic gardening) can deplete the soil so badly that the food grown upon it is not capable of sustaining life. Deficient soil produces crops that lose their properties for creating protein and mineral supplying products—food grown on such soil becomes empty vegetation with very little nutritive value. Why deprive ourselves of this all-important source of body-vitalizing elements? Why defeat nature's plan to contribute healthful fuel for human consumption (this too is an intrinsic part of the cycle of life)? The nutritional value of a fruit or vegetable—mineral, carbohydrate and the necessary vitamin content—is inextricably linked to the content of the soil itself. Soil treated with chemical fertilizers, aside from being deficient, is imbalanced and impure—so it stands to reason that the food this soil produces will be equally imbalanced and unhealthy.

Growing and eating organic food is a major step towards realigning yourself with the benevolent and excellently constructed forces of nature. What your intellect begins to grasp as you familiarize yourself with the natural processes through gardening, your body will manifest as a result of eating the very food you've grown! Things begin to fall into place and the flow of life, and the good sense it represents, are felt to be very much a part of you (for after all it *is* you and you are it!) We keep reemphasizing this concept of natural, flowing cycles because it is the basis of organic gardening and healthy eating—an understanding that will begin to assist you on many other levels of existence. The ups and downs of natural processes, so infinitely right and wise, are the very things from which our everyday lives are constructed. Comprehending this will help you to feel that what before may have seemed abstract or illogical is indeed just what is ordained by the universe that surrounds you. Give it some serious thought. Thinking organically will set you on a path of mental, physical and spiritual learning that you will treasure for the rest of your life.

In essence, we are saying that one cannot divorce the beneficial physical aspects of good gardening and eating from the mental and spirit-oriented ones. They lead to one and the other just as the processes of nature give rise to each other.

In a later chapter we have gone into great detail about specific nutrients offered by organic vegetables. This will be a help to you in deciding just exactly what you'd like to grow and eat.

COMPOST & MULCHING

Organic gardening is not a complex concept to grasp. In fact, just like nature, it is eminently simple and logical—based on the truest form of reasonability—the reason of natural flow and life-cycles.

There are two processes which make up the bulk of organic gardening work, and provide in their very definition a good understanding of the goals of organic work with the soil. These processes are composting and mulching, and both have for ingredients only natural organic substances—they employ no artificial or chemical substances whatsoever. Both composting and mulching are devised with the patterns of nature's own work in mind—and are designed to enhance and emulate this work.

If you can comprehend the basic principles of these two procedures you will have gone a long way toward understanding the very fiber of the organic gardening movement. It is a movement away from the chemical and the synthetic, and back to the wise and patient processes of nature. It is a realignment with the real core of living matter and life-flow processes, and, it is a move back to a saner way of life.

On the following pages we've explained the steps involved in composting and mulching and given information on why these two vital methods are so much a part of the organic way of life. And remember that they needn't be approached on a strictly large-scale basis to be tremendously fruitful and rewarding—you can utilize the idea of natural work with the soil no matter how small your own garden plot is going to be.

When we talk about returning organic matter to the soil, we are speaking basically of two different categories of organic matter. The first is raw organic material in the form of green leaves, stems or manure. Such raw matter can be added to the soil and do quite well in providing certain necessary nutrients. However, because it has not undergone the all-important process of decomposition it will not be as thorough in providing soil and plant nutrients as would already decayed matter. This is because many of the elements are still present in their more complex structure and are not immediately available for assimilation.

The second category of useful organic matter is compost. Composting is the decaying of organic matter. Organic matter in its raw state is not compost—raw organic materials are not available to plants as nutrients. Composting puts the organic elements through a process which makes the necessary nutrients readily available to the soil. Decay breaks organic matter down into its most basic components—and the compost heap possesses these very elemental substances.

Compost is the most valuable fertilizer available because it contains every vital nutrient the soil requires in the best form possible. To add compost to your soil is to add everything the soil and plants require in a state that they can use advantageously.

Compost is a natural plant food, constructed in the very way that nature nourishes plants herself. By utilizing the process of composting man can hasten this all-important natural function so that it takes place in a mere fraction of the time it would have normally taken. This is an excellent example of the supreme benefits of man's cooperation with nature, if he is but willing to learn her lesson and proceed with caution and understanding of the basic cycles he is dealing with.

By composting, we are short-cutting the life-cycle of plants, but still keeping things within the natural order. The materials employed in composting have traveled the way of natural early life and growth and the detour has come in by anticipating the normal process of death, to make these materials re-available to the cycle. This is to say that these substances are utilized before they would theoretically be available. What is important to remember is that ultimately, but at nature's much slower pace, the process of composting would take place. We are not intruding with any synthetic or unnatural process—we are just setting up a quicker schedule.

As things stand, it is not an exaggeration to state that compost and compost preparation are indeed the heart of organic gardening. The conditioning-fertilizing properties of compost cannot be surpassed!

The end result which makes composting such a worthwhile process is the formation of what is known as humus. The workings of decomposition as evidenced in the time-saving and marvelously efficient compost heap create humus, the most natural fertilizer there is.

The action of microorganisms within the compost heap (or within the ground when nature is overseeing this slower process) sets in motion the processes of decomposition which ultimately form humus. Being composted organic matter, humus is actually a combination of organic *and* inorganic compounds. This is because all organic material contains a partial amount of inorganic matter.

The decay of materials leading to the formation of humus is accomplished by processes which are microbial as well as chemical. Mineral nutrients, gases, and acids which will eventually be used in plant growth are liberated when various microorganisms exert their effect on the compost heap. These organisms include bacteria, fungi and algae as well as enzymes and yeast. The nature of these substances and their specific activities will be more thoroughly discussed in our chapter explaining the fertility chain. Simple compounds, easily assimilable for plant growth, are produced from more complex compounds through the processes of hydrolysis (compound decomposition by water), oxidation, and reduction. These chemical activities not only take place within the compost heap for decay of the substances in it, but also in the soil to which the compost is added. We have only to understand that the process of decay is in essence always accomplished in the same manner—regardless of where it is taking place. Once this process of decomposition has begun it will continue until all elements have been simplified to the point where they are reabsorbed for new plant growth. It is this phenomenon of continued, insistent action that makes the workings of nature so tremendously efficient.

The process of decaying as it continues hastens plant growth because decomposition gives plant life a better chance to use available nutrients. As soon as organic matter makes contact with the earth it begins to be transformed into its basic structures which are then in a form that can be utilized in the growing process of new plants. The cycle of decay is so all-pervasive that even a substance like humus, which by definition is the decomposed result of organic matter, itself decays. That is to say, humus is at once *decayed* and *decaying*. Eventually, in fact, humus grows old and its value to the soil diminishes significantly. It is for this reason that constant replenishment of compost is necessary and that the organic material content be continually renewed. The very essence of something living is that it moves unceasingly towards death—once again, this is the great ecology of nature! In creating a compost heap, we have actually isolated and stepped up a particular segment of the fertility chain—in a sense we are moving in on the "end" of the cycle and then using the results to perpetuate a new "beginning." But to understand nature is to realize that in truth there is no beginning and no end—just a constant flow, a circular, cyclical constant movement. And the main disadvantage of any kind of chemical, strictly man-made approach to gardening is that it ignores this crucial link between one natural process and the next.

The actual properties of the substance called humus are interesting ones. Because it is a combination of a vast amount of compounds (which are hardly ever the same from one compost heap to the next), it has characteristics which are quite difficult to pin down.

Basically, however, humus is most often of a very fine texture, although this is by no means a uniform law. It also has no real definite shape or structure, being what is known as an amorphous substance. Some of the compounds that make up humus are colloidal—this means that they are not water soluble, and even when mixed with water contain certain particles which remain suspended. Even though this is true, the countless nutrients available in humus are still readily absorbed by the plants to which it is applied.

The major elements present in humus are sulphur, iron, calcium, potash and phosphorus. There are also several trace (lesser) elements such as iodine, copper and zinc which are necessary to total plant health. Humus application is the safest and most efficient way to provide plants with these much needed minerals. Humus, along with these minerals, provides the kind of total organic environment which is necessary to the most beneficial use of the provided materials. A chemical fertilizer, providing certain isolated minerals, does so in a manner that is artificial and incomplete. The lack of other natural compounds makes the absorption of those that are present insufficient and lifeless. We'll go into more detail about the disadvantages of chemical fertilizers in a separate chapter devoted entirely to that subject.

Humus is not only the product of living material, but the source of organic matter as well. It is the *product* of living material because it is the final step in the life cycle of organic matter—the last form that matter takes before it is simplified and transformed into a component in new plant growth. Humus represents the final observable form that we can name in this process of decay. It is, at the same time, the *source* of life because it brings vital materials to the soil, contributing these necessary substances to the perpetuation of organic growth. And in its turn this new plant life will decay, once again become humus, and once again be broken down into basic structures that will foster more new growth. So what seems at first a paradoxical statement becomes just another observation in the cyclical, never-ending scheme of natural processes. When we begin to grasp this idea, and see its universal application, we begin to comprehend not only the natural processes of growth and the natural farming technique they suggest, but the very basis of all existence. Can you think of anything more fascinating?

Soil rich in humus has many advantageous properties which will insure healthy plant growth from fertile soil. Humus is instrumental in helping to maintain the spongy structure of the topsoil, without which the soil loses its vitality and ability to produce healthy plants. By contributing to the granular structure of the soil, humus assures excellent water absorption for the area (as the soil particles are held well apart) and also provides for maximum air penetration. Tightly packed, hardened soil can receive the benefits of neither air nor water—this is why maintaining a loose and spongy surface is so crucial. Without air and water none of the life processes can be carried on.

Humus in soil also contributes to the soil's ability to maintain a steady and beneficial temperature. Soil high in humus content will actually be warmer in the wintertime than soil lacking in this organic booster. Humus makes soil dark and rich in color. Dark soil absorbs heat from the sun more readily and retains it longer. This warmth stimulates seed germination and bacterial activity for more efficient decomposition. And the more efficient the rate of decomposition, the more productive will be the forces of life—all these activities being inextricably linked. To strengthen one process is to strengthen them all, though it is good to keep in mind that all functions must be carefully watched over to keep things balanced. Lack of balance is perhaps man's most destructive action in his chemical relation to the soil. The scales of nature, if we are to tamper with them at all, must be kept from tipping or the result will show in the quality of life produced. In other, oft-repeated words, we shall reap what we sow—and if our sowing is carelessly achieved, then what we reap will hardly be nature's best.

It is interesting to note that while the process of decomposition, by its very definition, renders nutrients more available for assimilation, the humus product can actually release elements that are normally locked in the soil, unavailable for plant consumption. We are not sure how this wonder is achieved, but by utilizing composting and the resulting humus we can avail ourselves of its benefits.

The high protein content of humus (about 35 percent) makes it quite valuable for the nitrogen requirements of plant life. A complex procedure which begins by the breaking down of this protein (present in the form of nitrogen) into organic matter and ends in the production of plant assimilable nitrates takes place readily in humus-rich soil.

WORK WITHIN THE COMPOST HEAP

Composting is based on the thorough decomposition of organic matter, and the fermentation necessary is highly dependent on the activities of the bacteria present in the heap. As soon as the compost heap has been constructed these organisms go to work— breaking down organic material through well-ordered and efficient actions. If this activity is carrying on as it should, there will be a great rise in temperature of the heap. In only a few days, a compost pile may record a temperature of over 160 degrees. Both a cause and an effect of the tremendous bacterial processes going on, this rise in temperature will also destroy any harmful weed growth that might be taking place within the compost. The rise in temperature is necessary to the proper timing of microbiological action. There are a host of different bacterias at work at different times—each doing its own thing at the proper moment to insure the maximum completeness of the process of decay. A good check on the activity of your compost heap is a temperature reading with a thermometer, just to make sure that the temperature is as high as it must be for proper decomposition activity.

GATHERING MATERIALS FOR A COMPOST HEAP

Collecting materials to begin your compost heap can be something you do with imagination and initiative. The size of your garden, and of course its location, will have much to do with the materials you decide to use.

For a start (and probably this will be all that an indoor or small home gardener will want to work with at first) organic kitchen wastes are convenient, practical *and* ecological materials for your compost heap. Instead of throwing out those old coffee grounds, potato peels, corn husks, vegetable parings and other things you have come to regard as useless garbage, why not turn waste into excellent nutrition for your organic plants? All these substances have tremendous life-giving qualities, indeed, they are quite alive themselves. It is one of the most sensible and beautiful aspects of organic farming that in the process of returning life to the earth, we are also ridding ourselves of the burden of disposal. Doesn't it make good sense to give these materials back to the soil where they can serve to replenish and refresh it? If composting were a national method of fertilizing our farmlands, we would not only be producing healthier, more vital food, we would be eliminating the pressing ecological problem of getting rid of wastes. Any alive, unprocessed food can be brought to the compost heap and utilized in the urgent cycle of life. In fact, most of the garbage you have available can be used to good effect in a compost heap. Just remember that any kind of soapy water or detergents are liable to contain harmful chemicals that will only harm your plants. It's also a good idea to avoid fat in your pile—it does not decompose well and will serve as an attraction for ants. Scraps, bones or any kind of other meat particles are welcome additions to the working compost heap.

Another easily obtainable compost ingredient is leaves. Rich in minerals, leaves make valuable fertilizer, but should be mixed with weeds and lawn clippings or they are likely to impede the proper aeration of the heap. Shredding leaves with a power mower makes them more readily assimilable and cuts down on their tendency to become tightly packed.

Sawdust is another excellent element for building the soil. It is easy to obtain from local saw mills and you'll probably be able to get it for free. Again, avoid tight packing because sawdust, too thickly and tightly layered, will restrict air circulation—it does not break down properly unless applied in fine sprinklings. This is true of sawdust in the process of mulching too.

Hay, especially if you can get it shredded, will decompose quickly and add a good deal of vital nitrogen to your compost heap. Like other green matter it is at its height of nitrogen wealth in its early stages of growth, and if you can get it during its first year, it will be all the better.

Tomato and squash vines, cornstalks, even flower stems are among the most valuable green matter you can lend to your compost heap. In fact, any plant remains will be a tremendous asset.

If you've been wondering what to do with the weeds you've been yanking out of your garden, why not give them back to the soil? Because weeds belong to varying plant families they are capable of extracting different elements from the soil, thus rendering a large array of vital nutrients available to your plants. This, of course, can only be achieved through proper composting. Don't worry about the seeds in the weeds, for the rising temperature in the compost heap will make growth of new weeds impossible. Therefore, the weeds you do have will be decayed and sent back in a basic and elemental form that will provide a great deal of nitrogen for the soil. Again, it is best to use weeds in their green form for full benefits of the nitrogen they contain.

Grass clippings can be saved for the compost heap too. Grass grown on healthy fertile soil to begin with, will, of course, be highest in the nutrients it can bring to your compost heap. For maximum nutrient value, it is best to cut the grass before blooming; you can let it wilt once cut, and it will be ready for quick decomposition when it hits the pile.

Back on the domestic front, nut shells can also be utilized in the compost heap. The speed of decay of each kind of shells varies with the particular kind of nut. Especially good are almond, pecan, filbert and walnut shells. Brazil nut shells and the shells of coconuts should only be used in a ground up form. Peanut hulls are tremendously rich in nitrogen—so don't leave them out.

For more vigorous and ambitious composting (which will usually be based upon your space and living conditions) there are several more rurally oriented substances for compost. Ideally, it is good to include at least some of these ingredients in your compost heap, but this of course is not always practical for the home gardener. Don't feel that your hesitation to use manure, for instance, need render your compost heap inferior; other high protein material (such as the meat scraps from your kitchen) is a fine substitute and any attempt at composting is far better than none. In fact the point we're trying to make in this book is that gardening needn't be a case of black and white, yes and no, all or nothing. Any organic matter you can put together will do wonders for your garden! We're including these next materials with the country gardener in mind. It'll also give you a clearer picture of the classic concept of full-fledged composting. We repeat, what is *desirable* in organic gardening is not a total *necessity* so don't be discouraged!

Manure is, to many people's minds, one of the basics of strict compost preparation. Its life-giving qualities exist in its possession of large quantities of nitrogen, phosphorus, potash and innumerable trace elements. Manure is valuable for inoculating the heap with tremendous numbers of plant and animal microorganisms which are vital to the initiation of quick decomposition. Manure is partly made up of small, undigested particles of food which have passed through the systems of animals and gone virtually unused. The estimation of microorganism life present in manure is somewhere around 20 percent. Like other "wastes" manure can be recycled for maximum soil fertility—and should be! The increase in the use of "clean" chemical fertilizers has not only created a great deal of manure to be disposed of, but has robbed the soil of one of its richest contributors. It is through the benefits of manure to the land that we see how animal life also comes into play in the master plan of nature. The cycle of life extends beyond the plant kingdom; plants nourish animals, thus becoming incorporated in their bodies. Eventually, through excretion, these life substances come back to the earth. To deprive the soil of manure is to cut it off from valuable returns in the process of give and take.

Along with the undigested particles in the manure are the digestive juices of the animal and a large population of bacteria which will contribute a great deal to the composting process. The manure of almost all domestic animals is high in nitrogen, phosphates and potash. The specific role that manure plays in the building of the compost heap will be discussed in our delineation of the Indore Method of Composting to follow.

By the way, small garden owners (and even indoor or roof gardeners) might like to experiment with dried manure, available at most fertilizer stores. For home gardeners who would like to use fresh manure, but don't have the livestock to produce it, a trip to a nearby stable with a small truck will probably furnish you with all you can use. And its' bound to be free!

Dried blood is collected in slaughterhouses and subsequently dried and ground. It is extremely high in nitrogen and phosphorus content and even a slight sprinkling of it in your compost heap will stimulate bacterial growth. It is preferable to moisten the dried blood before adding it, and once done it will just about guarantee a relatively fast-working heap.

Sewage sludge is a fine source of nitrogen and excellent for stimulating the work of the all-important bacteria.

Another excellent natural fertilizer and compost ingredient, which you should have no trouble getting at a local fertilizer store, is cottonseed meal. This meal is created when cottonseed is separated from lint and hulls and deprived of its oil. In a cake form, cottonseed is a valuable source of protein also used in animal feeding.

We've listed the major categories of composting materials you might want to consider using. There are many factors to consider in choosing what will be right for your garden in terms of space, time and effort. If you're growing a couple of plants in your living room and you've no space in which to isolate a compost heap, you can make use of the valuable kitchen wastes we've mentioned by sowing them into the soil. There they will decay perhaps a bit more slowly and be not as potent as an aged compost heap, but you're bound to reap the benefits of their life substances all the same. We'll talk more about the idea of indoor compost a bit later.

Now that you've got an idea of the materials you'll be using for your compost heap, you're no doubt concerned with the method of constructing it. As with all the facets of gardening we've pursued, this again is a question of degree and qualifying factors. A decayed mixture of organic matter is going to benefit your garden, no matter how you go about it. And there has been much careful research and experimentation done on the most expeditious and efficient ways to construct your pile. If you can apply yourself to this more ambitious process, then give it a go, by all means. And even if that's not possible, you'll still be able to observe and comprehend the basic concept of composting from your own humble heap.

Responsible for the most widely acclaimed and popularly used method of composting is organic farming's favorite hero, Sir Albert Howard. We've already mentioned some of Sir Albert's invaluable contributions to what we now know as organic farming, and his Indore Method of Composting is another feather in a very brilliant cap!

Sir Albert started out with the idea that composting is a crucial factor in successful gardening. Although compost has a history of loyal use dating back to the burgeoning farms of the Romans, Sir Albert Howard evolved a modern, *scientific* method of composting. Perhaps in using the word scientific here, it would behoove us to point out that science doesn't necessarily have to deal with chemicals and the latest synthetics. In fact, the dictionary definition of science is "systematized knowledge derived from observation, study and experimentation." In point of fact, science is *what works*, and the best science is the application of knowledge that yields the best results. Much of the outcry against organic farming is that it is old hat and unscientific! If the eminently efficient procedures of nature which have ruled the earth for eons (and despite much of man's presumption and disclaimers to the contrary still do and always will!) are old hat, perhaps these critics are right. And if scientific phenomena can be produced solely in laboratories with man-made substances, then they are right again. It should be remembered, however, that man has yet to create life, and while his ideas for substituting the "unscientific" processes of nature have seemingly speeded things up and given him the upper hand, the condition of our suffering earth tells a story of thoughtless destruction. If we are being harsh it is only because we feel an explosion of ecological awareness is long overdue! Ah, but we digress . . .

Sir Albert's Indore method has been recognized throughout the world as the most practical and productive means of fertilizing the soil known to man. His main emphasis was on the fact that composting needs to be accepted as a top priority of good farming

—that it is not a sideline to be dealt with lightly. Towards this end he devised the Indore method. (Indore, by the way, has no relation to composting in your house! Indore is the Indian state in which most of Sir Albert's work was carried out while he was an agricultural advisor there.)

An explanation of this method will provide you with some interesting insights into the dynamics of the compost heap, and even if you are unable to imitate it exactly, may give you some clues toward making your own attempt a successful one.

The basic idea of the Indore method is that the compost stack be built in layers. A recommended width of six or seven feet assures that the materials in the heap will not dry out and will remain moist enough to stimulate bacterial action. The heap can be set slightly underground or built on top of the ground with or without a wooden enclosure.

But, regardless of how the heap is laid out on the ground, it is best to begin with a matting surface of brush and/or coarse twigs. This has a two-fold purpose. Air will be able to escape through this matting far better than if the heap were constructed flat on the ground. This underneath layer will also permit excess water in the heap to drain out. Both these functions are necessary for a well-balanced and well-functioning compost heap.

On top of the matting should be first a layer of *varied* green matter. Using one kind of grass clippings or other vegetation will tend to provide the finished heap with an ability to furnish only one type of nutrient. There is a basic rule in composting that the more diversified the ingredients, the more beneficial the final product.

The green matter layer can be bolstered with fallen leaves and weeds along with the grass clippings and garden residues you might have on hand. This layer should be about six inches thick and it should be kept in mind that tight packing is a deterrent to good compost progress; in fact it is pretty much a rule of good gardening that tightly packed soil or organic matter is likely to give you trouble due to its general impenetrability. It's also best to avoid very coarse materials for this layer; the coarse matter is good for the matting, but shredded, finer vegetation is best for the first layer of green matter. Large, fibrous pieces of organic matter are slower and less efficient in their process of decomposition.

The second layer of the Indore heap is made with fairly clean manure. Again, it's a good idea to use a mixture of manures—that is, from several different types of animals to continue the ethic of organic diversification. This manure layer is not only high in nitrogen, but it is literally alive with the high number of micro-organisms that will spur on the decomposition process, sustain it, and later strengthen the microbes in the soil to which the compost is added. There are other vital nutrients available in the manure of animals but the most important asset is the microbe content.

It is advisable to use fresh manure whenever possible, but if this is not feasible, dried manure (which is easier to handle) can be utilized instead. At the same time it is important to realize that older manure (having aged prior to being composted, that is) just might furnish a humus that is not as rich in vitality, nitrogen and general potency as compost from fresh manure would be. It is also likely to slow down the composting process. The manure layer should be two inches thick.

The third layer of compost is the soil layer, and it is here, with careful consideration, that you can add the extra minerals and nitrates you think your soil might need (unless you are going in for pretty serious gardening, you are probably not going to spend the time assessing specific soil content). But no matter what you think your soil might need, it is necessary to sprinkle lime or phosphate rock in with the soil. The lime will do much to counteract the extreme acidity released through the processes of organic disintegration.

The procedure from now on in is a repetition of this basic pattern. First, six inches of diversified vegetation followed by a two-inch layer of manure (if you're using poultry manure, one inch here will do). Next comes the one- or two-inch soil layer mixed with lime and any other desired minerals. As the stack is built, each layer should be soaked with water—naturally the dry layers will require a more thorough dousing than the green ones.

The heap should be built approximately five feet in height, and should not be tamped to excess. Part of the natural composting process is the settling of the heap to its proper stature. The final layer should be topped with soil, hay or straw and left concave to catch rain. A way to insure ventilation is to put posts in the heap down through the middle (about 4 inches in diameter) starting after the first layer. When the stack is finished these poles can be removed and the holes they leave will provide the necessary ventilation.

By Sir Albert's calculations the whole process takes about 12 weeks, but the heap should be turned after about six, making sure to place the top and side material into the middle, so that it too may decompose at maximum speed. After 12 weeks, the stack can be turned once again and then will be ready for use on the land.

There you have it—a process which is simple in design but is, in all likelihood, beyond what you are planning for your garden (at least in the beginning). If you're unable to include all these varied ingredients in your stack, just try to use a diversification of what you do find available. One ingredient compost stacks are very rarely successful.

If you are planning to go the Indore compost route, its' a good idea to begin work on it when you're done with the garden production of the preceding season. October and November are excellent times for beginning a stack because plant materials are available in great number. If you do begin in late fall, you'll be sure to have compost when spring gardening time comes along.

As for placement of the compost heap, there are no stringent rules. In an outdoor garden you shouldn't have too much trouble locating it in a convenient, and not distractingly prominent place. If the dimensions of Sir Albert's heap are too large for your small backyard, then by all means experiment with a reduction in dimensions. You have nothing to lose by it, and you're bound to learn lots. Traditional composting indoors is not terribly practical, but if you think you'd like to give it a try, you can attempt composting in a steel drum or wooden bin. The impracticality of indoor composting lies in the need for air. You can't compost in a vacuum or anything even vaguely approaching one. So if your garden is an indoor proposition, perhaps you can get a friend to let you use a portion of his backyard for your composting. (Admittedly, this brings up a problem of transporting the finished product, but where there's a will there's a way!) Another possibility is constructing a container for compost on the roof of your apartment building.

The New Zealand compost bin is constructed from two-by-fours forming a rugged frame, with a lattice of boards nailed over it. Because air circulation is all important, the top and bottom of the bin must be left open.

The steel drum compost bin can be used, provided it is at least 6 inches off the ground (this can be accomplished by setting it on a circular metal frame with legs).

In any case, if you are attempting strictly indoor gardening, it is probably best to collect your kitchen wastes (perhaps accumulating them in a closed container for sanitary purposes if you wish) and plan to work them directly into the soil. Their organic components will still do wonders for your plants, even if they haven't been composted in the traditional way. We are making do perhaps, but we are still growing organic vegetables and that's what counts! More on the subject of indoor composting in our Indoor Gardening chapter.

USING THE COMPOST

Once finished, the ideal time to apply compost is a month or so before the actual planting of your garden. The method of application is quite simple. After thoroughly stirring or turning the soil, the compost should be added to the top four inches. Compost added for vegetable and flower growing is best utilized if first put through a ½-inch sieve. In a small garden, this should not represent too much of a problem. If you're adding organic material that hasn't been composted, it's a good idea to break it up as thoroughly as possible, either putting it through a grinder or shredding it manually. This refining of the compost makes nutrient absorption all the quicker.

Compost should be applied in amounts of one to three inches per year. Unlike chemical fertilizers, you needn't worry about putting in too much. The application can be made once or twice yearly, depending on your assessment of how badly it is needed. Just keep in mind that the more you replenish the soil with organic matter the richer it will be in vital life-giving substances. But too-frequent application will not benefit (the soil can absorb only so much) the plants either, so try and keep it on a reasonable level. You'll find with organic gardening in general that your own common sense and intuition will play an important role. And the reason you can trust your senses is because you are working along with nature, and as far removed from her as we all feel sometimes, we are still very intrinsically involved in all these processes ourselves!

Any way you slice it, composting and the life-giving humus that results from that endeavor are at the very core of the organic gardening concept. In the process of composting we can observe the workings of nature at her best, aided by man at his best—man in a calm and rational state of land management. To realize that we must constantly be returning to the soil those elements which we demand of it, is to comprehend not only the necessity for composting, but the overall scheme of organic farming. Having considered the simplicity and crystal-clearness of this way of aiding and abetting nature, it is difficult to understand why man has decided to rape our planet. Unfortunately, the age in which we live is not as conducive to rational, natural thinking as it is to the lure of the quick dollar. Without attempting to give a lecture on morality or psychology, we have to say that man is being taught his lesson now that the facts regarding the extent of the damage he has done are being made known. It is impossible to discuss the myriad, rewarding benefits of organic gardening without making frequent reference to the alternative of chemical, synthetic farming that the majority of this country's farmers have chosen. We can only regard with growing dismay and concern the destructive results of the work done in chemical farming by Justus von Liebig. His work in and of itself was certainly laudable, it is only the extremes to which his discoveries have ben applied that we bemoan!

There are many excellent and instructive books available on the subject of organic farming compost methods. We suggest: COMPOSTING: SANITARY DISPOSAL AND RECLAMATION OF ORGANIC WASTES, H. H. Gottas, *World Health Organization*; THE COMPLETE BOOK OF COMPOSTING, Rodale, *Rodale Books*; MODERN HUMUS FARMING, Sykes & Friend, *Faber and Faber*; OUR SYNTHETIC ENVIRONMENT, Lewis Herber, *Knopf*; HUMUS AND THE FARMER, Sykes & Friend, *Faber and Faber*; FARMING AND GARDENING FOR HEALTH OR DISEASE, Sir Albert Howard, *Faber and Faber*; HUMUS: ORIGIN, CHEMICAL COMPOSITION AND IMPORTANCE TO NATURE, Selman A. Waksman, *Williams & Wilkins*; FERTILITY FARMING, F. Newman Turner, *Faber and Faber*; THE LIVING SOIL, Lady Eve B. Balfour, *Faber and Faber*.

MULCHING

The other natural process which organic gardeners swear by is mulching.

Mulching is the laying down of a blanket of organic materials over the soil in which crops have their roots. A thorough job of mulching includes covering all the surrounding soil and permitting only the plants themselves to go uncovered. We can see a first-rate example of this in the manner in which nature covers a forest floor.

Like composting, mulching serves a multitude of important purposes. A layer of mulch is an invaluable aid in helping to conserve the moisture of an area of soil. It is also excellent for maintaining a fairly even temperature; a good strong layer of mulch can actually protect plant root systems from the devastation of a sudden frost. Mulch acts as an insulating blanket, keeping crops warm in winter and cooler in the summer, thus avoiding extremes. Acting in a fashion similar to compost, mulch adds minerals and nutrients to the soil as it decomposes. Of course, this process is much slower when you're dealing with mulch, because the elements are not previously decayed. Like compost, mulch is first rate at preserving the spongy quality of the topsoil, thus insuring good air and water circulation.

A layer of mulch on the topsoil of your garden will cut down tremendously on the amount of weeds that crop up. Most weeds will not be able to push their way through a solid layer of mulch, and the ones that do will be easily uprooted.

It goes without saying that this particular benefit of mulch is both time saving and back saving for the gardener who otherwise might have spent hours a day at the tedious business of weeding.

Just as it protects roots from the excesses of heat and cold, mulch also keeps the sun and wind from penetrating the soil too deeply and evaporating the moisture that the plants so desperately need. It is also a deterrent to erosion, one of the soil's most destructive enemy processes. Soil underneath a layer of mulch remains cool and damp—perhaps the most beneficial condition for topsoil that will stay resistant to any of the hostile activities of our often unpredictable weather.

Last, but certainly not least, is mulch's ability to act as a bruise-saving cushioner for ripened vegetables as they become heavy and begin to sag towards the ground. In the same way it prevents the mildewing and rotting of vegetables which would ordinarily touch the ground and become moldy as a result of the contact. Mulch keeps vegetables clean and quite dry.

It is wise to keep in mind that mulching can be a detriment if not properly tended to. There are a few basic pitfalls to keep in mind when you're planning to set about this process.

Seedlings which have been set into very moist soil should not be immediately mulched. Organic matter which encourages a high humidity of the soil can also create the damping-off condition in young, unestablished plants. Damping-off is a plant disease which is almost 90 percent fatal and occurs when over-moist soil becomes infested with a kind of harmful fungus. Therefore it is a good policy to allow seedlings to get a firmer "foothold" in their soil before doing extensive mulching. You can, however, mulch the larger surrounding areas, as long as you are careful to leave clear a good-sized area around the base of the plants. Another reason for avoiding close contact with plants in their early stages is that mulch in its decomposing process claims a good deal of nitrogen. Young plants cannot compete and are likely to go into a state of nitrogen deprivation if they are forced to by the proximity of mulch. When a root system has enlarged itself and is not strongly dependent on the soil at its base, it is safe to bring the mulch close around the plant stem. At this point the plants are not deprived if some of their base nitrogen is utilized by the mulch surrounding them, and they will profit from the beneficial aspects of the mulch.

Mulching should not be undertaken immediately after a very heavy rain, because the soil is waterlogged and if mulched will be the perfect victim for an onslaught of harmful fungi. And when you do mulch, make sure that mulch substances such as peat moss, manure and ground compost do not touch the base of those plants. The soil will need to remain open to the air so it can dry sufficiently.

It is not difficult to avoid mistakes in mulching if you keep in mind the fact that some of the beneficial aspects of this process can prove detrimental to plants too young and weak to stand up for themselves, and to soil areas that have undergone a severe conflict with the elements. Obtaining materials for mulching will not give the home gardener much trouble at all. Almost any organic waste material can be used for mulching. We have listed the various types of mulch you may employ and how they can be most beneficially utilized.

Grass Clippings

Excellent for use in composting or mulching, grass has a reasonably high nitrogen content. Grass is often used as a mulch for flower gardens, but can certainly be used with good results in a vegetable garden as well.

Paper

Yes! Paper! The benefits of paper as a mulch are threefold. Set out in a vegetable garden and covered over with a layer of straw or woodchips to insure that it doesn't blow away, newspaper and magazine paper are highly efficient at discouraging the growth of weeds. It is also dense enough to keep rain from passing through to the soil, a fact which should be kept in mind as it may become a danger in an overly dry area. Four to six thicknesses of paper (best applied after a heavy rain) will protect the soil and at the same time add humus to the soil by its own process of decomposition.

Leaves

Also valuable in the form of compost or mulch, leaves have a way of matting and sealing in moisture, at the same time being a valuable source of humus and mineral materials such as calcium, magnesium, phosphorus and potassium as well as nitrogen. They are quick to decay and readily give the nutritional benefits that process engenders to the soil.

Stones

One of the most natural and widely evident types of mulches, stones have a few additional advantages that other mulches don't enjoy. Rain and snow bring rock minerals in contact with the soil, thus creating a valuable natural fertilizer. The conditions beneath rocks and stones on the earth are ideal for the work and growth of the helpful bacteria and other organisms so crucial to the perpetuation of the fertility chain. They also provide an ideal environment for the activities of the earthworm. Rocks give an added boost of warmth to the soil, especially in seasons when healthy plant growth needs this assistance.

Sawdust

If sawdust and wood clippings are to be used as mulch they should be well rotted. Failing this, it is a good idea to mix the sawdust with a few shredded leaves or bits of straw to allow for aeration. Sawdust is best and most effectively applied when plants are about two inches high. At this point a one-inch layer of sawdust can be spread over the area, but should not be deeply worked into the soil. Sawdust in immediate contact with the soil will begin to decompose right away. Both sawdust and wood chips are excellent soil improvers, but wood chips with their higher percentage of bark naturally have a higher nutritive value to the soil. Plentiful quantities of wood chips are becoming available throughout the country as more and more farmers and gardeners begin to use this valuable mulching substance.

Other mulching materials which will do the job admirably are peat moss, pine needles, weeds, packing materials, cornstalks and even the dust from vacuum cleaners.

The value of mulching remains undisputed among organic farmers today, and is gaining an enthusiastic following among farmers who don't subscribe to all the tenets of organic growing.

As far as the amount of mulch necessary for maximum benefits goes, it is a good idea to make sure that during the growing season the thickness of the mulch should be sufficient to prevent the growth of weeds. Finely shredded materials do a more efficient job of this and less of them are required than when using the unshredded and looser varieties of mulch.

Ruth Stout, the nation's foremost advocate of year-round mulching, uses a system almost totally dependent on spoiled hay. This is what is known as a technique of "permanent mulching" and years of faithful experiments have proven Mrs. Stout's convictions to be well founded. This spoiled hay is perhaps the most practical method for the home gardener. It is easily peeled off from its "books" and ready for application that is convenient and free of mess.

Other excellent sources of mulch for home gardeners with a small garden and not much access to the more rural forms of mulch are: coffee grounds, cocoa and buckwheat hulls, shredded cotton burs, peanut shells, and spent hops, the waste product of breweries. Hops are also valuable in discouraging some insect pests.

Mulching is a process very much in agreement with the ways of nature. It is the very process by which forest land is protected and fertilized. Like composting, mulching is an effort to keep gardens free of synthetic, unnatural substances; to constantly return to the earth that which has been taken away and to do this in the sanest, most organized way possible. Mulching is a clean and simple process with benefits that are overwhelming considering how uncomplicated a method it really is. If we continue to study the basics of organic gardening and realize that it is a constant effort at repeating and enhancing the natural processes of the earth, we will see that these are excellent ways of dealing with man's need to perpetuate the produce our earth can bear.

Both mulching and composting are at the very heart of the organic farming movement. More and more farmers are becoming aware of the excellent results the use of these two methods offer. This is primarily because nature "approves" of the mulching and composting procedures. Indeed, the ideas for both methods were devised from nature's own plan.

Organic matter is the basis for both composting and mulching. They are simply different approaches to the use of these vital substances. Mulch is primarily a protective "blanket" for your garden—it is useful for weed prevention as well as guarding some of your crops against frost. The organic matter applied as mulch does not necessarily have to be decayed first—and it is best used in its fibrous state. Composting, of course, involves the process of decomposition prior to adding the organic matter to the soil, and works most effectively when the decayed matter is in a fine-textured state.

For more thoughts on mulching, its uses and benefits, try the following: HOW TO HAVE A GREEN THUMB WITHOUT AN ACHING BACK, Ruth Stout, *Exposition*; PLOWMAN'S

FOLLY, Edward H. Faulkner, *University of Oklahoma Press*; GARDENING WITHOUT WORK FOR THE AGING, THE BUSY AND THE INDOLENT, Ruth Stout, *Devin-Adair*; GARDENING WITHOUT POISONS, Beatrice Trum Hunter, *Berkeley Medallion*; THE BASIC BOOK OF ORGANIC GARDENING, Robert Rodale, *Ballantine*.

These books will provide you with a broader view of the mulching process and guide you in your work with this most valuable of organic gardening methods.

CHEMICAL FERTILIZATION

We mentioned, in our description of modern organic farming's beginnings, the work of Justus von Liebig in the early nineteenth century. Liebig's work was truly genius in many respects, but his one unforgivable mistake was that he neglected to view the soil as a living component in the universal scheme of life that surrounds us on every side. A laboratory chemist rather than a farmer, von Liebig did not have a feel for the land; he did not understand emotionally or spiritually the essence of life as it bursts forth from the earth. He didn't comprehend the ever-flowing cycle of life we've been laying before you in this book. And, in not seeing the constant dance of life and death, ebb and flow, dark and light, within the soil, he could not appreciate the gravity of the error he made.

Von Liebig's process for determining the elements needed by a plant was to burn the plant and then chemically analyze the substances that he found in the ashes. It was these substances, then, remaining after the plant had been burned, which were necessary for the plant's growth. As sound as this conclusion may appear, it leaves out one thing which is the very basis of plant life upon the earth. That factor was that along with the elements evident in von Liebig's experiments, plants need a host of other major and minor elements—all of which are a part of the basic organic scheme of life. He did not realize that simply maintaining the mineral content of the soil would not suffice, but that a careful watching over of the organic content of the soil was crucial to a healthy plant's environment. It has been repeatedly proven that disregard for any ingredient in *any* aspect of the life cycle will ultimately result in an impoverishment of the entire system. To persist in returning only mineral nourishment (and chemical nourishment at that!) to the soil, without attention to providing a continuing supply of organic elements, has led to the situation of imbalance and poor soil health that has robbed this country *of over half of its topsoil* in 200 years. In considering the impact of this, we should also keep in mind that it takes from 500 to 1,000 years to replace one single

31

inch of topsoil. We have lost, in 200 years, somewhere between three and six thousand years of the earth's work to create a fertile topsoil.

People are beginning to sit up and take notice of the condition of our land. They are beginning to question what has caused the rampant deterioration of our crop-producing soil. It is our contention (and the contention of organic gardeners everywhere) that the widespread use of chemical fertilizers and the general disregard for the natural processes of good land management are certainly to blame for our dying soil. By furnishing the land with incomplete, imbalanced commercial nourishment, we have seriously interrupted and impeded the workings of nature. Now is the time to stop this destruction and take steps to restore our land to its former glory and fertility. This can be accomplished only through organic farming and the land management concepts that go along with it.

We have already mentioned the failure of chemical farming to regard plant growth in terms of whole systems. It is this type of one-sided thinking and action that is responsible for the tremendous ecological imbalances today. This is the kind of thinking that Sir Albert Howard referred to as "fragmentation"—a way of regarding soil in its mineral or chemical sense only. It is this concept that is at the top of the list of the passionate arguments against chemical fertilization—that it is only one, small piece of nature's reality that cannot function independently of other considerations equally important. This blindness to the life cycles we keep talking about is the uppermost objection to commercial synthetic farming that organic farmers are expressing.

In considering just how serious these dealings with chemical fertilizers are in our country, it is relevant to note that the majority of farmers in the United States are in many cases totally dependent on these commercial fertilizers for use on their land. One type of chemical fertilizer, the kind supplying nitrogen, enjoys a national yearly usage of over 7 million tons! The fragmentation of which Sir Albert spoke has become in great part a way of life for today's farmers. Most of them have no real idea of the natural relationships between the chemical properties of soil and the organic matter present in it. Use of chemical fertilizers has drastically cut down on the numbers of farm animals kept by modern farmers. Feeling no use for the "old fashioned" method of manure fertilizing, they have given up keeping animals and turned to the handier, more convenient chemical substitutes. Thus, today's farms lack the vital balance between crop and animal life. The essential role that manure can play in producing hearty, healthy soil is ignored by these farmers. Regarding manure as troublesome waste (in the processes of nature there is no waste!) they phase out their animals, thinking that they've evolved a modern and commendable method of farming. What they have done in actuality is taken a dangerous shortcut that sees its results in the steady decline of our soil's fertility and the accompanying decline of the quality of food produced from that soil. We can observe another cycle here, though now it is not a happy chain of events: disregard of the balances of nature leads to land deterioration; land deterioration in turn leads to produce that is steadily losing its life-giving, nutritional qualities; food that has ceased to be vital and life supporting leads to a deterioration in the health of an entire population. Even in our very neglect of nature's laws, we come full circle back to this cyclical reality.

Chemical fertilizers have a tendency to bring about unnatural changes in soil make-up that seriously disturb the workings of the great benefactors of the soil. Very often these disrupting properties of chemical fertilizers are due to the artificial method of making all chemical nutrients water soluble. In nature, soluble food elements occur very rarely. We already have discussed the colloidal property of humus, which somehow manages to nourish the soil immeasurably, despite its insoluble structure. To make these minerals (naturally insoluble in their genuine state) soluble, man has added various acids and other chemical substances. This artificial method of making minerals available has a way of changing the structure of the soil the fertilizers are added to. With years of continued use these processing chemicals seriously unbalance the soil's supply of nutrients. So, ironically enough, what man has added to enhance soil nutrition and fertility serves ultimately to throw the soil's entire regime off balance. Many of these chemicals are not only unnecessary, but poisonous, and can therefore only bring destruction to the land, the produce and the consumer of that produce. This subtle and gradual quality of poisoning of the land was in good part responsible for the instant and unquestioning acceptance of chemical farming. It took a few years before the detrimental side-effects were evident. By that time the industry had gotten going in a big way, and there was no stopping the crushing, thoughtless wheel of farming "progress."

The onslaught of chemical destruction of soil is truly insidious —but it is no longer a secret, and yet still it continues!

The chemical system of gardening and the avid way it has been pursued since its inception are grave testimony to man's vanity in attempting to outdo nature. On the other hand, the working with organic material and natural elements, which are the very core of soil structure, is a hopeful symbol of the almost lyrical cooperation which can come to pass between man and nature.

Land treated exclusively with chemical fertilizers becomes sterile—an environment that is not capable of handling the constantly busy, very much alive workers of the soil. Earthworms will eventually die out of a soil that is so treated and in order to keep any worm population at all, they must be reintroduced time after time. Even then it is difficult to keep these important soil cultivators from becoming extinct in such an area.

At the same time, microbes, algae, and fungi will not survive and continue their work in a sterile, artificially treated soil. In order for them to be present in the soil, it must have some degree of organic content. Soil that has not had its organic content replenished will not be able to support these tiny organisms. This is a double threat at least—the organisms' presence serves as a barometer of soil fertility. If they cannot survive in the soil it is a good (or rather bad) sign that the soil will not bear crops worth eating, if indeed it bears at all. And any hope that a depleted soil might have had for restoring itself through the efforts of the earthworm and microorganisms is defeated because they cannot remain long enough to effect a change. The life force of these workers is strong, but they cannot stand up to the devastation of chemical fertilization for long.

Soil that cannot support life within its own structure will certainly not be able to fulfill the life-force needs of human beings. On the other hand, we have seen how the application of organic matter, whether it be fully composted into humus or only partially so, will immediately improve the health and population of these soil benefactors. To withhold this nourishing and all-important booster from the soil is to condemn it to a slow and painful death.

The new and dangerous compounds which are often formed by repeated treatment of soil with chemical fertilizers have another characteristic. Aside from generally creating chaos in the usually well-ordered system of soil nutrition, these compounds "lock up" food substances in the soil in a way that makes it utterly impossible for plants to assimilate them. And we are talking about substances that are exceedingly important in plant growth. In our discussion of humus, we mentioned a characteristic diametrically opposed to this destructive one; humus has the ability to render previously unavailable nutrients ready for use by the vegetation. On the one hand we have an interference with nutrients that are normally accessible with some new substances that were unavailable before the addition of humus. The choice is ours to make!

Many of the chemical fertilizers that chemists construct feature compounds which are so strong they can actually penetrate to the subsoil workshop and disrupt the work carried on down there. This may be evidenced in the sudden and unnatural, unhealthy growth of a plant or in a destruction of the biological agents which safeguard the permanent good health of fertile soil. Some of these products are strong enough to bypass the feeder roots which protect the innards of the plant, and enter the plant anyway, giving a disagreeable taste to the vegetation that is apparent even to livestock. Organic materials act as a buffer against over-powerful intrusions of any kind, so it is the chemical farmer who needs an organic additive all the more. The fact being that the only way to use chemical fertilizers without serious soil/crop damage is to strengthen it with repeated applications of organic matter. And of course the very farmers who are using the chemicals so faithfully at the same time refuse the method of organic buffering. They are committing errors of omission as well as commission, and it is the land that suffers. There are some farmers who subscribe to the use of commercial mineral fertilizers as supplements. They improve and strengthen the soil through organic methods and use caution when making the supplements with these artificial substances. While there are multitudes of organic farmers who would insist that this is certainly not kosher, still it is a saner and more balanced way of treating the subject of chemical fertilization.

In providing plant life with nitrogen, phosphorus and potassium only, as is the scheme of chemical farming, a group of other elements is being completely left out. Recent research has uncovered the fact that trace elements, although they make up only one percent of the plant's needs, are still exceedingly important in the healthy and vigorous development of vegetation. In turn, plants grown in soil without these trace elements can only provide food that is deficient. It has been found that many of our diseases today are the result of a deficiency of trace elements in our diet. They are easily obtained from plants grown in rich and complete soil, but are not available from produce grown in artificially treated soil. This is because they can only be released for plant consumption through microbiological organism activity. In the case of chemical fertilizers this activity is prevented and so the trace elements such as zinc, iron, iodine, copper and cobalt remain locked up and unavailable to crops and, ultimately, to humans. Lands that have been cultivated using organic methods are high in the type of microorganism activity necessary to liberate these trace elements. In fact, organically treated land is likely to provide more of these substances than the plant is able to utilize. And, in adding organic materials to the soil for the sake of liberating these trace elements, the impetus for the release of assimilable portions to the plants is also taking place. Organic matter added to the soil provides the raw material as well as the necessary processes to make the raw material available where it is needed most.

The chemical fertilizer people did not overlook the discovery of necessary trace elements to plant growth. On the contrary, once word of the need for these substances was out, fertilizer manufacturers began to develop "added extras" and "power boosters" supposedly meant to supply the missing trace elements. They took the artificial substitute process one step further in doing this. The major danger in using these boosters is that it is almost impossible to gauge what would constitute a safe amount of chemical trace elements. By their very nature, trace elements are present in such minute quantities that it is not possible to imitate these exact proportions in mixing chemical fertilizer. And there is a very definite danger in adding too much of a given trace element; an overdose" of any of these substances will render the soil toxic not only to the plants that grow on it, but to any animals that eat these plants. Trace elements are persistent and difficult to get rid of; in other words, they stay around for a long time once they are introduced into the soil. They don't wash out, even in a heavy rain, and it is nearly impossible to check them once they are present in improper proportion. This careless and haphazard method of adding trace elements to the soil is seriously harmful to all facets of the life cycle—the soil, the plants and the animals. Once again, chemical farming has defeated its own purpose—in an attempt to create more abundant plant growth, it has poisoned the soil and contributed to the eventual deterioration of the life processes with which it comes in contact. Plants may seem to flourish and grow splendidly for a time, once injected with these synthetic super chemicals, but what seems like success is not only short lived but highly destructive. As our soil dies, we move on to new land that has gone unsullied. It is only a matter of time until these new fields are in their turn destroyed and rendered useless for growth. The concern of ecologists (and indeed any thoughtful citizen) is that ultimately there will be no more land to cultivate. Along this rather frightening line of thought, it has been estimated that the disappearance of soil and the accompanying disappearance of all plant life on earth would mean the extinction of all animal life within the period of a year. This is perhaps the best expression of just how dependent man and animal are on the healthy, natural functionings of the life cycles of soil and plants. There is no way to circumvent these cycles or to outsmart them—they are intrinsic, essential and far-reaching; man can disrupt them but he cannot prevent the ultimate ensuing devastation—it is bound to be brought home to us eventually. After 200 years, we already can witness a vast amount of the destruction we have wrought.

Another irony in the strictly chemical approach to land fertilization is that soil which is in poor shape due to an insufficient sponge structure cannot improve this crucial layer unless it has organic material added to it. On any plot of extremely depleted soil the most urgent need is the replacement of a healthy sponge structure. Without one, in a soil that does not possess a good measure of organic materials, there will be no ability for absorption. In this case, the addition of the chemical fertilizers to a sandy soil not bound together by sponge is almost futile because the applied ingredients will trickle straight through. In the case of a lacking sponge structure resulting in the opposite condition (a tightly packed soil) fertilizing substances will not be able to *penetrate* the soil. For these reasons, soil which has lost its total spongy quality is hopeless. Hopeless, that is, unless the gardener or farmer works to establish a new sponge structure. He will certainly never achieve this through the use of chemical fertilizers alone. Again, we can see a continuing thread of illogic that allows man to make only partial restitution for the good elements he takes out of the soil each year.

Nitrogenous fertilizers have a tendency to create acid conditions within the soil. At the same time, experiments have proven that one of the quickest and most complete methods for ridding the soil of earthworms is to add a type of superphosphate fertilizer to the land. In Australia, *nine-foot long* earthworms, originally present in a vast and hearty number, were completely eliminated after application of this type of fertilizer—and, such nitrogenous chemical substances are equally harmful to other beneficial soil organisms.

Having made what we consider to be a rather persuasive and reasonable case against the use of chemical fertilizers, we are hoping that their destructive, nature-disturbing qualities have convinced you that this method of trying to outdo nature has been the undoing of man! Chemical interference has seriously altered nature's processes which, undeterred, could have continued in their own way to insure the productivity and healthiness of our soil. It is this drastic interloping on the part of man that has necessitated the most vigorous efforts possible from organic farmers—to restore the aching, desolate soil. Unfortunately, the continuing (and enthusiastic!) addition of such chemicals to our land is becoming a more real and dangerous threat as time goes on. To begin to cultivate your soil naturally, no matter how small and insignificant a plot you think you have, is to do your own part in helping to revitalize our dying earth. And the more people who do their own individual share, the more hope there is that we may somehow salvage not only this basic life structure, but the entire process of natural existence. These are indeed somber thoughts to think on, but your personal contribution to the growing refusal to add more chemicals to the land will bring you a feeling of satisfaction and joy.

For further information on the properties and results of chemi-

cal fertilizer usage, check these other excellent sources: OR-
GANIC GARDENING AND FARMING, Joseph Cocannouer;
FREAKS AND MARVELS OF INSECT LIFE, Harold Bastin,
A. A. Wyn; SOIL DEVELOPMENT, Edward H. Faulkner, *Uni-
versity of Oklahoma Press*; GUARDIANS OF THE SOIL, Joseph
Cocannouer; OUR SYNTHETIC ENVIRONMENT, Lewis
Herber, *Knopf*.

THE DANGERS OF CHEMICAL PESTICIDES

It was in the late 1930s that the insecticidal properties of the chlorinated hydrocarbons (DDT and several others not as well known) were first discovered. Spurred on by emergency measures engendered by the needs of war, the use of these materials was accelerated and greatly expanded, primarily in the 1940s. This was the beginning of the use of "hard" pesticides. Previous to the discovery and exploitation of these substances, farmers had employed a variety of more natural means. One of these was a system known as biological control, a concept based on the idea of using living natural enemies of a pest organism to reduce its population. Garden and farm pests were an age-old problem, and man, even pre-technological man, had evolved several time-proven, traditional means of coping with the problem before the chemists stepped in and paved the way for large-scale production of artificial pesticides.

The general effect of strong chemical pesticides on the cycles of nature corresponds in many ways to the effects of man-made fertilizers. They are both guilty of a fragmented, partial and disruptive view of the way of coping with plant problems. Both chemical fertilizers and pesticides fail to respond to the many-faceted requirements of nature. Because it is an ongoing, ever-changing, many-sided process, the cycle of life must be dealt with (if indeed we "deal" with it at all) in a manner that makes provision for its highly diverse and all-inclusive character. Just as chemical fertilizers emphasize only one aspect of the soil's need by restoring only basic mineral elements, so do chemical pesticides emphasize an unremitting destruction of insect life without other, necessary considerations. The basic problem is one of oversimplification; chemical pesticide use does not provide for the fact that there are many species within the plant-animal community which are beneficial to plant growth and must be permitted to survive. Strong chemical insecticides, while they may be applied to deter only a certain type of harmful insect, act also to kill or suppress a great many more. This immediately destroys the highly effective system of biological control by affecting the survival of competitors, parasites and natural enemies, the existence of which is the very basis of natural insect control. The result of this nondiscriminative insect destruction is a harmful alteration and dislocation of the entire plant-animal community—another example by throwing the processes of nature into chaos and imbalance. By suddenly and drastically removing whole populations of insects, regardless of their positive or negative role in the health of the plant and soil, all links in the chain of organic life are adversely affected. We can see the analogy between this danger and the disruption of the fertility chain of the soil by chemical fertilizers. Basic to the very conception of these man-made substances is the fact that they are one-directional and insensitive to the real needs of the earth. If man is looking for dramatic and instant results in terms of destruction of insect populations, then he has certainly found the perfect tool in these pesticides. Unfortunately, the accompanying (and equally drastic) side effects have long-term, severely damaging results on all of life.

When we speak of pesticides, we are talking not just about insect-killing chemicals, but also of substances that destroy any kind of supposedly harmful life in horticultural or agricultural areas. This general category includes not only insecticides but fungicides and weed killers which share the propensity of destroying beneficial organic life as they kill the actual pests.

There are other devastating factors in the use of chemical pesticides, even aside from their tendency to disrupt natural balance. Because of their highly toxic nature and the persistence of these substances, they tend to infiltrate and remain in the tissues of living things. As this poison is carried from the simple portions of the natural food chain into the more complex and higher areas, the concentration of poison increases rather than diminishes. These accumulated poisons reach high proportions, and rather than diminishing or disappearing by the time the food is cooked in our homes and served on the table, they are magnified. The idea that cooking or washing will destroy the toxicity of these chemical-permeated foods is simply wishful thinking and, at worst, propaganda designed to placate the concerned consumer. Continuing and intensive research on the effect of chemical pesticidal sprays and powders on the human body shows that once taken into the system, the poisons are stored in the fatty tissues. Such toxicity has been found in mother's milk—in fact the proportion of DDT found in breast milk in laboratory studies is actually four times the amount permitted to be present in publicly sold milk. There are residues of the poisons discovered in stillborn or unborn babies, and DDT is capable of doubling the rate of human mutation, according to Dr. Osny Fahmy, a British geneticist. Cancer research has established the fact that there is a firm link between that disease and chemical residues in our food. Even with all this alarming information available to the public, the outcry against use of these pesticides has not been nearly strong enough to result in the discontinuation of their use by farmers. On the contrary, the use of these chemicals and the money poured into that industry continue to skyrocket. Chemists and scientists whose valuable time and energies could be applied to the study of safe and ecological ways of controlling insect pests are instead devoting their time to producing stronger (and hence more lethal) pesticidal substances. Behind this dismaying fact is industry—just as in the case of the exploitation of von Liebig's work with chemical fertilization, industry has stepped in and made chemical pesticides a highly lucrative business. The powerful people and corporations behind this successful business venture are not about to give up the high profits they are gleaning. The consumer is being duped in a most unforgivable and frightening manner—and only with a vast escalation of public outrage and expression of that outrage will this dangerous charade be brought to an end. It is not true that without chemical pesticides this land cannot produce healthy, vital crops or that organic and natural methods of insect and weed control are outdated and passé. On the contrary, as long as farmers employ the use of these highly toxic and poisonous sprays, we cannot hope to have healthy, life-giving, unsullied food at our tables. In fact, it has been estimated that even if DDT were never used again, it would take more than another hundred years for the substance to be totally gone from our food. As things stand now, every mouthful of food we eat in the next century will contain at least one molecule of DDT. The persistence of these chemicals has allowed them not only to enter the vital food-producing chains of nature, but to become resident within the very soil.

In terms of sane and logical thinking, man has no choice but to immediately discontinue all use of DDT and similar dangerous pesticides. But at present it seems the best we can hope for is the growing move toward organic farming and a healthy way of growing that includes replenishment of the soil with organic matter. Certainly overnight results will not be had, but the course is charted and though it's a long hard road back to healthy soil, we have taken a giant step when we begin to farm and garden organically.

The reason that technology is striving to produce stronger and more potent pesticides every year is that pests quickly build up an immunity to the chemical compounds with which they are attacked. It works like this: after their use, chemical sprays create an insect vacuum for a period of time, but when the pests return they are present and active in approximately 15 times the proportion in which they were originally evidenced. This, of course, calls for the invention and preparation of an even stronger agent which acts in the same way and must eventually be put aside for the next, more potent (and more destructive) chemical killer. The average chemical insect pesticide is capable of killing off somewhere between 100 and 1,000 different insect species—and as we've said, this is with no regard to the true role of these different species in the life cycle. Indeed, the use of chemical pesticides creates a cycle all its own—a constantly escalating cycle of misuse and abuse of life processes. Because nature is so totally and irrevocably cyclical, any dealings with her, even if they be drastically inappropriate, will cause the firing of a chain of events. In the case of pesticides it is a chain of indiscriminate death and jeopardy for everything that lives.

IN SUMMARY

Before chemical pesticides became widely used in the 1940s, farmers used natural, traditional methods including biological control. Such "hard" substances interfere with the cycles of nature by acting to destroy helpful as well as harmful species of organisms. While these chemicals act quickly and thoroughly, their detrimental side effects are tremendously destructive to all of nature.

The DDT which is permeating the soil from which we harvest

our vegetable and fruit produce is being carried from the field to the table and into our own systems. The effect of DDT as it works its way through the various planes of life processes becomes more and more injurious to the health of humanity. Research has shown that molecules of DDT have severely destructive effects on many of the organs within the human body and that much disease today can be traced, in part, to the use of these chemicals in growing our food.

Chemical pesticides may wipe out an insect population speedily and completely but it has been proven that the insects will ultimately return in even greater numbers and eventually build up an immunity to each type of chemical that is administered to them. For this reason, much time and money are being spent discovering and manufacturing stronger and more toxic pesticides. The chemical approach to control of garden pests is a merry-go-round of ever-increasing intensity and proportion that will lead only to more extensive soil, plant and health deterioration.

We've only scratched the surface of this highly controversial topic. For further reading we refer you to: GARDENING WITHOUT POISONS, Beatrice Trum Hunter, *Berkeley Medallion*; INTRODUCTION TO ORGANIC GARDENING, Chuck Pendergast, *Nash*; HOW TO GROW VEGETABLES AND FRUITS BY THE ORGANIC METHOD, J. I. Rodale, *Rodale Books*; FERTILITY FARMING, F. Newman Turner, *Faber and Faber*; THE BASIC BOOK OF ORGANIC FARMING, Robert Rodale, *Ballantine*.

In our next chapter, we've outlined for you some alternative possibilities to the use of poison in your home garden.

SAFE INSECTICIDES;
A RATIONAL ALTERNATIVE TO POISON

We have spoken at length about the dangers and myriad ill effects of employing chemicals as deterrents to insects and various harmful weeds and fungi. By now, you are well aware of the need to abandon chemical gardening practices and return to the use of more natural methods of pest control. As we mentioned, prior to the invention and widespread use of chemical agents in farming, man had very definite and successful ways of coping with the problem. There is little doubt that using organic methods of pest control can be a bit more time consuming at first, but when you comprehend the benefits derived from this approach you will agree that it is time and energy well spent.

The organic view of pest control is, like every other aspect of organic gardening, an overall observation of the totality of nature. It takes into consideration *all* the various interrelated processes that go on around us—and in dealing with pests, organic gardening is careful not to throw these processes into a state of chaos. For too long now, man has sought to dominate nature rather than to work along with her. And the main ethic of organic methods is just this kind of cooperation. We assess the problem—we observe nature's own way of resolving it—and we seek to emulate and enhance those very methods. Once we accept this idea of working with, rather than against, nature, we will have the key to successful perpetuation of life and growth.

Biological Control

The idea of using natural enemies to reduce pest population is a concept that goes back to pre-historic days. The first real scientific work with biological control, however, was carried out in the late nineteenth century. At this time, scientists were applying the idea of biological control only to the reduction of living pest organisms by other living things. Researchers began to work with specific pest problems and specific natural enemies. At that time the California citrus industry was being threatened by the destructive

45

work of a small insect that fed on the sap of citrus leaves and twigs. There was no known remedy for the insect plague and trees all over the state were being gradually destroyed. In the meantime, a man named Albert Koebele began to study the problem seriously —working with the idea that there might be a natural enemy of the insect which could be introduced into the area in order to destroy the insect population that was killing off the citrus trees. His work led him to the discovery of just such an organism. This was the vedalia lady beetle which preyed on the very insect responsible for the citrus calamities! A shipment of these beetles into the California region (where they thrived, by the way) resulted in a successful controlling of the destructive insects. Within two years, they were under control. The lady beetles continued to thrive and feed on the insect pests, thus keeping the California citrus crop alive and well. This is a perfect example of the marvelously complementary workings of biological control. And such control required no poisons, no toxic sprays, no disruption of nature's processes. The vedalia lady beetles were not dangerous to the citrus crop or to the beneficial insect life in the area—they simply destroyed the detrimental organisms that were threatening the fruit.

We have said that one of the major malevolent characteristics of chemical pesticides is their indiscriminate death dealing to all the life around them. We can readily see, then, how biological control circumvents this dangerous situation, while at the same time coping quite efficiently with the problem at hand. In the specific instance of the California citrus crop, all went well until the mid-1940s. The introduction of widespread chemical insecticides at that time was responsible for the decimation of the valuable vedalia lady beetle. Gardeners and farmers who were enthused about the idea of complete and speey insect extermination didn't realize that they would also be killing off one of their crop's most prized protectors. Indeed, after the application of chemicals to control another type of insect, the lady vedalia died out and the insects they had been killing flourished once again. Had these farmers taken the time to locate another natural enemy for the new type of insect, they could have avoided rendering the beetle's biological control inactive. In fact, it is the introduction of chemical insecticides which has ruined the many successfully functioning biological control systems which preceded them. Chemical fertilizer and pesticide use is just not compatible with the workings of biological control. And this again is due to the fact that chemicals cannot be programmed to destroy only certain segments of the insect population—they take a very large and a very general toll. They interrupt life processes and interfere with all levels of working organisms within the soil. And they destroy the hope of controlling plant pests naturally and safely.

This early work with biological control was limited to setting up insect enemy relationships. As the years passed, scientists began to consider the possibility that this same system of natural control could be successfully applied to establishing resistant plant strains through breeding, as well as some more expanded work with insect populations. Everything was based on the understanding that insects and plants were vitally woven into each other's life activities, and that work with one type of organism could have important bearing on the life and health of the other. This is indeed the very essence of organic living and ecology—to make use of the relationships of all living things to each other and to their environment. None of the intrinsic life processes are thrown out of whack when *natural* relationships are experimented with—there is no introduction of foreign substances to the atmosphere.

Biological control is very much at the core of organic work with the soil. It is a natural method for controlling life upon this planet, and as such, it is bound to be both efficient and beneficial in the long run. But, unfortunately, the growing emphasis on chemical insect control has not only destroyed many of the necessary agents in biological control—it has taken many researchers out of the field of natural pesticide work and set them to experimenting only with chemicals. Too often, the chemical approach does not allow for the possible virtues of natural processes. Having become too exclusively involved with synthetic substances, researchers will not spend time working with new modes of natural control which they could be developing.

Biological control is at work within the environment and the gardener or farmer may not even be aware of it. Natural enemies are quietly keeping insect pest populations to a minimum so that their destructive effect on crops is negligible. In many cases, application of chemicals to the soil has killed off the insect enemies and caused the formerly harmless insects to swell to a highly destructive number. At this time the work of biological control, natural and plodding, becomes evident. This can occur to the extent that the newly increased insect pests become more danger- ous than the population that the chemical was originally engaged to destroy. Chemical treatment of the soil is filled with many ironies —and this situation is certainly one of them. A plan that originally seemed quick and convenient (i.e. destroying insect pests with a wave of the chemical wand) results in an unexpected catastrophe. This is because man has left the scales of life unbalanced, and in nature, there is simply no way to get away with it! This balance has been based on the fact that competitive forces continually equalize insect and animal populations—to play disrespectfully with this balance is to bring havoc down on all the systems that function within the natural world.

Some other work within the realm of biological control involves using microbes, or insect viruses to control destructive strains of insect pests. These viruses function as natural enemies, killing *one and only one* type of insect. They will not be harmful to bene- ficial insect life and will have no effect on any other organisms. This method is known as bacterial control and at present there are 280 insect viruses isolated and capable of doing this work. Bacterial control is different than chemical control action in that it takes a few days for the pest strain to die off completely. Unfor- tunately, these non-instant results are often too much of a trial of modern-day gardeners' patience. What actually happens is that the insect becomes inactive and harmless just a few hours after bacterial contact—it just takes a while for the pest to die com- pletely. To many people this is not as reassuring as the immediate effects of strong pesticides which cause anything that's susceptible to their poison to curl up and die on the spot.

Another type of natural control is achieved with the use of "juvenile hormones." Application of these hormones to insect larvae will prevent the insects from growing into adults and repro- ducing. Such substances are applied in microscopic amounts and have no negative effect on any organisms other than the insects they are used to stunt.

The beauty of biological control lies in its ability to cope spe- cifically with plant pests. It can zero in on a destructive population and wipe it out or reduce it to a harmless minimum. It is also nature's own basic scheme for keeping life balances intact and as such it cannot be improved on by man, but merely perpetuated and enhanced. Home gardeners who are not interested in biological control and feel they are handling a potential situation of pest invasion can obtain natural control substances such as the bacteria and hormones from biological control centers throughout the country. Robert Rodale's highly informative chapter on this subject contains a listing of where such substances are available, which you will find in his THE BASIC BOOK OF ORGANIC GARDENING, published by Ballantine.

Indoor gardeners who suspect that there might be a serious insect problem (and this is really quite unlikely) will want to locate some of these substances—there are a few other alternatives for the outdoor gardener, however.

One of your most useful friends in your organic gardening venture is the bird population of your area. There are more than 50 species of birds in this country that can be drawn to your garden with their very persistent appetites for certain insects which are tasty to them, and destructive to you! There are scores of methods for attracting birds to your land. You can provide them with materials for nest building and suitable places in which to build their nests—this will keep them close to your property. It's also a good idea to give them water to insure that they remain happily in your area when you need them most. You can also attract birds with various food supplements such as bread crumbs or kitchen crumbs, suet or dried fruits. The cultivation of various types of plants such as marigolds, poppies or asters will attract birds of desirable species to your garden, too. Once again, THE BASIC BOOK OF ORGANIC GARDENING by Robert Rodale will provide you with valuable specific information on attracting the birds of your choice. It is good to remember that not all birds are going to be helpful, and that you would do well to choose wisely. There is also the possibility that you'll feel your garden doesn't require an escalated population of the feathered variety—and that's fine, too.

There are several non-toxic sprays which can be used by the organic farmer as insecticides. It is always preferable to resort to these types rather than the commercial sprays—even if you can locate commercial varieties that are fairly non-toxic. These may be applied in a pinch, but it is best to work with more natural, "homemade" remedies that you can trust completely. For your information and experimentation we've listed some of these time-tested insecticides that will not damage your crop or your land.

Ryania

This is a perfectly natural insecticide which is derived from the roots of ryania speciosa—a plant from South America. Ryania is very useful against corn borers, codling moth, cotton bollworm and several other pest species. It works by sickening the pests rather than killing them—and in so doing renders them harmless by causing them to lose their appetites! Using ryania is not as good as using no insecticide at all—but if you do have an insect problem that needs some kind of attention—you can use ryania with far less concern than any store-bought chemical insecticide. Some gardeners are fortunate enough not to have to employ *any* external insecticides. But this is in gardens that have maintained a firm and healthy balance—a balance that is attainable through continued use of organic gardening methods.

Rotenone

Rotenone is an insecticide derived from a combination of tropical plants and it can be used safely on plants to kill harmful insects as well as external parasites of animals. This plant derivative is a contact and stomach poison which will affect insects, but is of very low toxicity to man or animal. Rotenone has the disadvantage of being a short-lived protective agent—that is, it has very little residual effect. You should also keep in mind that insecticides such as this are not as selective in their effects as is biological control through natural enemies, bacterial action or hormone application. But if you do need to use a general insecticide, it's better to turn to rotenone than a synthetic compound that is likely to be toxic. There is a plant called Devil's Shoestring which contains 5 percent rotenone and may be used toward the same end.

You can use plants as insecticides without utilizing commercial products at all—you can achieve this by growing certain plants with insecticidal properties right in your garden. If you'd like to attempt this, try planting some marigolds or chrysanthemums, or any aster family plant in your garden. There is a good chance that they will do much to keep insects away and you won't even have to bother applying anything to your garden—just simply plant them and allow them to grow—they'll do the rest. There are also many herbs which are useful and effective as insect preventatives —among these are tansy, anise and coriander. You can grow delightful herbs in or around your garden and have the added advantage of reaping their pest-deterrent qualities.

There are several other substances—concoctions, actually—that will aid you in your battle against insect pests, or with any detrimental organism that is disturbing your garden. One of these is the juice of sweet peppers. Sweet pepper juice is effective in a battle against certain virus organisms. Such juice in various research tests proved to successfully inhibit the growth of a variety of destructive plant viruses. It is also possible to combine a variety of vegetable juices of hardy plants and spray this mixture on ailing plants. This has the effect of altering something in the faltering plant that will render it more resistant to virus activity. Cultivating resistance is one of the basic principles of organic protection against plant pests of all kinds. By strengthening vegetative growth through organic matter additions and through these various natural sprays, organic methods improve the general plant constitution. Just like the human state of health, plants remain firm against disease when they are initially hardier. Strong plants resist not only bacterial and fungous disease, but insect invasion with greater effectiveness. You can chop up or grind plants which seem healthy to you— onions as well as peppers are excellent for this purpose—and dissolve the substances in water. With such a solution you can bathe your other plants, and witness a strong effect on ants, spiders and worms of the harmful variety.

A very simple and useful treatment for plants suffering from fungous disease is a plain hot-water bath. The heat involved will destroy fungus without injuring the plant—especially valuable with a fruit crop. There are also experiments going on today with an eye toward using hot *air* to rid fruits and vegetables of decay fungus.

Did you ever hear of fighting plant virus with milk? Well, it's being done on an increasingly large scale by organic gardeners and farmers in this country. Milk diluted with nine parts water and sprayed on plants at two-week intervals has been found exceedingly effective in ridding the crop of harmful viruses. You can actually use the milk right out of your refrigerator for this purpose, if you have reason to believe that your garden is significantly troubled by viral infection. It's just one more way of working with the problem naturally—and avoiding the use of toxic substances that are liable to infect the soil, the plant, and ultimately the bodies of the members of your family who consume the food you grow.

Nature has her own system of checks and counterchecks. Another of these is the concept of companion planting. You can work with companion planting by mixing your vegetable rows with different types of vegetables and even a sprinkling of flowers. This mingling of crops often serves a multitude of purposes, and our chapter on "What to Plant" will offer some useful suggestions in this direction.

In Summary

Anyone who's growing a vegetable crop on his "home turf" will want to make sure that he is doing it as healthfully as he possibly can. Outdoor gardeners will want to make sure that they are not disturbing the many-dimensioned life cycles within their soil. No citizen who is concerned with bettering our environment will want to use harmful chemicals at any stage of his home gardening process. And the indoor gardener will certainly be wary about introducing any chemical substances into the air of his living room! When we realize how chemical agents used in large outdoor gardens contribute to the pollution of our air, it is unthinkable to consider polluting the air we live, eat and sleep in, with chemical toxins and odors. And to help the organic gardener in working without such poisons, we have attempted to familiarize you with the alternatives. Biological control is perhaps the single most comprehensive pest prevention scheme of nature and we would do well to utilize its highly effective concepts.

We've also outlined to you the natural products you can purchase that are in tune with the biological control method. These are in the form of either bacterial substances or "juvenile hormones" which you can use without fear of abusing either your soil or the plants. You can also use these agents with the secure knowledge that the food you eventually gather from your garden will be safe for your family's consumption.

In addition to the types of insect and general pest control available with biological control, there are various types of sprays and baths for your plants. These include pepper juice sprays, onion and water sprays, milk and water baths, and plain hot water baths. These methods can be used with complete confidence in your home—indoors and out.

You can also use the ground-up leaves and roots of healthy plants because these plants are resistant to whatever pests—insect or bacterial—might be disturbing your crop. Among the most potent of plant insecticides is rotenone—which is sometimes called derris and is a product of certain tropical plants. Rotenone is available in pure and unadulterated form from your pet shop or veterinarian. Remember that it is often mixed with synthetic, toxic substances in its commercial form—and these should be carefully avoided. Another plant product is ryania, a powdered substance derived from a South American plant. But keep in mind with any plant-derived insecticide that they are alien to the natural control level of your garden's make-up and even though quite non-toxic, are best avoided.

Companion planting is a most natural and rewarding way to control pests within your garden. Check our chapter on specific vegetable planting for the best way to go about companion planting.

There you have it—all the various alternatives to poison pest control that the organic gardener can work and experiment with. These are all excellent methods for making sure that your garden produce and your home environment remain as pure and clean as possible. This is home ecology at its best and most practical—and everyone concerned with making our earth a better place to live will want to give the traditional as well as the more modern natural methods a try.

For further and more specific reading on this fascinating subject, we suggest the following books: GARDENING WITHOUT POISONS, Beatrice Trum Hunter, *A Berkeley Medallion Book*; THE PRINCIPLES OF BIOLOGICAL CONTROL, Harvey L. Sweetman, *William C. Brown*, as well as Sweetman's book THE BIOLOGICAL CONTROL OF INSECT PESTS, *Comstock*. Also see A HISTORY OF THE VEGETABLE KINGDOM, William Rhind, *Blackie & Sons*; INSECT MICROBIOLOGY, Edward A.

Steinhaus, *Comstock*. The ambitious reader will also find scores of informative treatises in back issues of United States Department of Agriculture journals as well as gardening journals of all descriptions. If you're interested in further, more comprehensive works on the subject, consult the subject card catalog at your local library under "Biological Control" or any related topic you're interested in. It is indeed an intriguing study and gardeners interested in the wondrous natural balances of nature will be eager to find out more. This learning process can be carried out not only by reading, but by practical experiment in your very own home garden!

UNDERSTANDING THE SOIL

Now that we've given you some of the basic whys and where-fores of organic gardening and also done our damndest to convince you that all the pleasure of food growing and the delights of delicious eating are not restricted to people with spacious country homes, we're ready to descend on you with the how to's. And to lead you into the method of gardening, first a discussion on the very basis of all growing things—the soil.

When contemplating the idea of just exactly what soil is, it is good to remember that it represents one of the crucial links in the chain of natural elements. These elements are soil, water, and organic and inorganic plant life.

Our good earth has not always been covered with the soil we see upon it today. In fact, there was a time when the entire planet was a mass of solid rock with no vegetation whatsoever. But, within this rock existed primitive forms of fungi and bacteria. It was actually these organisms and their physical activities which worked to liberate certain acids, both inorganic and organic as well as carbon dioxide in the rock. In turn, these substances slowly caused the rock to break down. Through this process the formerly solid rock became (ever so slowly!) a substance capable of producing further forms of life. In this process we can witness the precision of nature's cyclical timing. Certain elements give rise to other elements which in their own good time contribute to a final result. The way of nature is slow, granted, but it is highly efficient and admirably organized. The point of organic gardening is to work maturely to hasten these natural processes, not by circumventing or ignoring them (as does chemical farming) but by giving them a wise and helping hand.

Also contributing to the process of creating soil from the original solid rock were the forces of heat and cold, wind and water—as well as the effects of massive glaciers exerting tremendous pressure against the rock structure. As the rock surface was worn away, thus becoming increasingly penetrable, mosses and various types of lichens began to protrude. These lower order types of vege-

55

tation, in their turn, died and contributed their remains to the surface they had sprung from, enriching it and leading to the growth of higher forms of plant life. As this process continued, both simple and more complex forms of vegetation began to cover the earth. As animal life evolved, these living organisms contributed first their wastes and eventually their bodily remains to the soil—also thickening it and rendering it more fertile for plant growth. As time passed the soil became increasingly rich and capable of producing more robust and complicated forms of vegetation.

In this manner evolved the earth's surface as we know it today. To best comprehend the structure of soil for a clearer understanding of how the growth process takes place within it, it is best to visualize a layered earth. The uppermost layer is what we call topsoil, and constitutes that portion most directly involved with gardening. This area is the principal feeding zone of plants although plant roots do connect to lower areas or subsoils, where there exists a good deal of mineral nutrition for them.

The topsoil bears a sort of vegetative mulch on its surface, which aids in preventing a crusting-over of the earth's surface. Immediately beneath this vegetation is a kind of spongy structure which serves to connect the topsoil with the lower layers of the subsoil areas. Acting as a kind of network, the spongy layer sends elements which result from the decaying of surface vegetation down to the subsoil. This lower area acts as a kind of storehouse and workshop where these elements are held and added to, eventually to be sent back up to the topsoil as nourishment. The process by which soil nourishment takes place is dependent on a host of different organic substances known as the fertility chain. This fertility chain is made up of a group of various "links" which are each, in their own way and for different reasons, a vital part of the natural harmony necessary for the production and maintenance of healthy, vital soil. Each "link" does its very own job towards a common end—a design which is the ingenious creation of nature.

LINKS IN THE CHAIN: FUNGI & ALGAE

The plant life which functions totally within the soil's structure is composed of fungi and the small green plants known as algae. Perhaps the most important type of fungi to the health of the soil is the mold. It was not until the last century or so that molds began to be regarded as friends of the soil. Research has now revealed that harmful soil fungus is insignificant compared to the amount of beneficial types of mold.

Specifically we have mycorhiza, which is actually the association of a kind of fungus with the roots of a plant—a partnership from which both members benefit. This name is generally applied to the type of fungus which lives in this manner and thus assists the plant as well as its own organism. There is some link between mycorhiza and penicillin, though this important drug is not technically of the mycorhiza type.

Mycorhiza works by entwining its hyphae around the rootlets of the plant; a situation which at first sight might seem harmful for the latter. Through the excellent research of Sir Albert Howard it was learned that these mycorhiza threads were actually carrying food materials directly into the plant! And even more beneficial, the mycorhizae break down the food substances to make assimilation in the plant easier and more complete. At the same time the mold itself is profiting, thus forming what is known as a symbiotic relationship—a scheme of nature in which both the mold and the host plant reap valuable benefits.

Another important asset of the mold mycorhiza is its ability to provide a plant with much-needed phosphorus—even in soils where this element is weak or wanting. Micorhizae are able to accumulate phosphorus and then feed it directly to the plant, once again in a readily assimilable form. There is further evidence that this valuable mold has the talent of acting as a "fixer." "Fixing" being the process of taking essential elements from the air and processing them for use by complex forms of vegetation. Sir Albert Howard contended that mycorhiza could perform this feat with both nitrogen and phosphorus—two of the major elements in plant growth.

The presence of a good amount of organic matter in the soil insures the growth of these valuable mycorhiza molds. When it is provided with a good, spongy environment in which to work, mycorhiza can improve the work of other important soil agents. The hyphae of mycorhiza, extending into the upper regions of the soil will serve as food for any living substance that can reach them. Soil with a high level of mycorhizae activity is bound to be healthy

—acting at once as a barometer of fertility and as a contributor to that same quotient of vitality. Excellent for spurring on the growth of mycorhiza is the application of compost and the accompanying enriching qualities of its humus product.

It should be remembered that a good part of soil health is determined by maintaining the sponginess of the topsoil. Porous, loosely packed soil has the ability to circulate air and water in a way that hardened soil cannot. Compost added to the soil will help maintain this spongy quality. Cooperation with the soil is a tremendous aid to the growing process—for soil is only alive in the sense that it can harbor growth. When we speak of "living" soil we mean specifically a soil that is fertile and "alive"—able to foster vital plant growth. Because soil cannot renew itself at a rate that is beneficial to the needs of man, continued addition of organic material will be a valuable boost to all the links of the fertility chain.

ALGAE

Although not quite as important as the mycorhiza molds, the algae within the soil are an indispensable link in the fertility chain. Acting as tiny food factories, algae are capable of producing a good deal of soil nourishing substances, especially the protein compounds. The elements produced in the soil by algae are used by other members of the fertility chain for *their* nourishment. The algae provide essential substances to the molds as well as soil microbes, a beneficial type of soil bacteria we shall discuss directly. As these workers of the soil join forces in the fertility chain, they build processes for the world of higher plants—thus contributing to the rhythmic flow of nature. The processes of death and decay are also a crucial part of the scheme; when algae die they become absorbed by higher plants in the form of protein. In nature there is no haste—but ah! no waste!

MICROBES

It is necessary that a plant be rooted in soil that can help to provide it with the protein all of its growing parts require. Because a plant must manufacture the largest percentage of its own food by itself, the fertility chain in the soil should be prepared to provide the plant with everything it requires in the proper proportions. Another necessary tiny organism in this process is the microbe.

Microbes are actual soil germs present in the millions, responsible for much of the life-giving activity that goes on in the soil workshop. These beneficial bacteria control practically all of the chemical changes that take place in the subsoil region. In speaking of the subsoil workshop it is necessary to make an important distinction between food and food substances. We can understand it thusly: created in the subsoil workshop within the earth are the food substances that nurture the plant so that it may in turn produce (through photosynthesis) a finished food product, edible for animals and humans. There is a continuing interdependence between the food the soil offers the plant and the food the plant offers us! This reasoning is what is behind the organic gardener's insistence on healthy, vital soil—no nourishing, life-giving substance ever came from a hardened, depleted soil.

One of the most important functions of the germs and bacteria in the soil workshop is to guard the feeder roots (leading from the workshop to the leaves of the plant), to insure that only desirable, nutritious substances enter the plant—and they usually do a thorough job of it. Would that we had such efficient police to watch over the things we send into *our* bodies!

Microbes exist in several different varieties within the soil—but as with the algae and mold, work with an incredible harmony as a single unit to help sustain healthy plant life. And as the different types of microbes work together they also contribute to the general workings of the other links in the fertility chain. More of nature's beautiful and practical rhythms!

Microbes do their most important work by bringing about the decay of organic matter. They are a crucial part of the decay process which in its own way is as necessary to the life cycle as any of the so-called living processes—indeed these two processes are inextricable factors in the same grand plan. Germs aid in the disintegration of organic matter which has ceased to live. In the process of falling apart, these decaying substances return most of the elements they drew from soil and air back to those sources. Once again we can witness the minimum waste of life materials and energies that nature allows. This process of microbe-assisted decay is the combined and harmonious effort of the various kinds of microbes present in healthy soil. When they have done their work, only organic compounds or free elements remain. This overall scheme is known as the Nitrogen Cycle.

The process by which nitrogen compounds are altered according to nature's master plan is one of the most basic of her functionings. As it exists in the organic world, nitrogen may appear as protein. When this nitrogenous molecule of protein has ceased to live, it disintegrates, and in the process of doing so returns its nitrogen to the environment. The cycle is one of complex nitrogen compounds being returned to their original simple state. In this simple state, nitrogen is once again free floating. It is here that the microbes come in. Acting as nitrogen "fixers" these germs can take the denitrified (free) substances and make them once again available for use in plant growth. And so the cycle begins again and again and again.

The microbe's ability to act as a "fixer" extends also to sulphur and phosphorus. With their important contribution to the nitrogen receiving status of the plant as well as the sulphur and phosphorus materials, we can see that microbes are essential to the process of building as well as the function of decay. As with every other part of this exciting and efficient fertility chain, microbes are vitally interrelated to all the processes of growth. To isolate one process from the next is to miss the whole point of the cyclical essence of the fertility chain.

Present in even the most depleted soil, microbes are always ready to respond to an application of the proper kind of material— those living substances found in compost and manures. To give a craving soil these substances is an important step in restoring the fertility chain which will ultimately lead to the restoration of growing power. This is achieved directly through the buoying up of the faltering microbe population. This is based cn the all-important improvement of the spongy vegetation surface of the soil—guaranteed to escalate the multiplication of microbes and, in turn, give a significant push to the work they so efficiently accomplish.

We have only to create a conducive atmosphere for the microbes and they will do the rest in maintaining an active and flourishing soil. To the extent the gardener is willing to work at providing organic material for the soil, so will the microbes respond and do their part.

EARTHWORMS

Is it difficult to believe that a creature such as a headless, eyeless worm could be responsible for performing some of the most vital functions necessary to healthy plant growth? Perhaps it is, but it is nonetheless true that earthworms are an *absolute necessity* to the existence of fertile, "living" soil, and that these tiny beings are capable of a host of marvelously beneficent favors to Mother Earth.

Earthworms actually eat their way through the earth. As they move through the layers of dirt they digest a tremendous amount of soil, passing it out in a form which is richer than its original pre-digested form. These castings have been scientifically proven to contain elements highly valuable to soil nourishment and subsequent plant growth. An earthworm can digest its own weight in soil in a matter of hours, passing it all out and moving on to fertilize even more. For this very reason, Aristotle called the earthworm the "intestines" of the soil. During their continuing journey through the soil, earthworms manage to churn the layers of earth in a way man or machine cannot duplicate. This process helps to improve the soil's structure, loosening it and thus providing for improved water and air circulation so vital to a plant's healthy growth.

In essence then, the earthworm eats the soil, digests it and in that process conditions it. Burrowing into the ground as far as six feet down, they break up hardpans which have been created by chemical fertilizers and other artificial horticultural practices. The earthworm-burrowed holes act as actual watering tubes for the soil, increasing the rate of rain absorption.

The earthworm also draws leaves and other valuable green matter below the surface of the soil, which in turn decays and nourishes the plant roots in a manner they would not have enjoyed without the help of this friendly little creature.

After a life that has been an invaluable boon to the health of the soil and plants with which they come in contact, earthworms die and in doing so provide perhaps their greatest contribution to the land. Their decaying bodies, which contain an oil with a very high nitrogen content, become fertilizer within the earth. Since the average earthworm life span is a year, or two at the most, the large number of decaying earthworm bodies furnish a substantial amount of this all-natural fertilizer to the earth.

Because organic matter and mineral rock fragments are the natural food of earthworms, the addition of these elements to the soil will insure a healthy and burgeoning population of these unsurpassed soil helpers. But strong chemical fertilizers and other artificial substances create conditions which are highly distasteful to the earthworm population. We have already pointed out the sure-fire worm-exterminating properties of chemical fertilizers. One such substance, in particular, ammonium sulphate, is particularly harmful to the earthworm. The earthworm population, which we should be working to increase, is slowly but surely dying off through the use of chemical fertilizers. Potent insect sprays are another enemy of the earthworm.

To help boost the earthworm population in a given plot of land, inoculations of earthworm eggs or capsules can be utilized. Once the earthworm community has gotten a good start it will grow and flourish if not disturbed by the harmful interference of chemical additives. To provide the organic matter which the worm craves as food is to further assist the continued work of these wonder worms. By nurturing the worm population, you are providing yourself with one of the most valuable single assets of soil cultivation.

SOIL MINERALS

Nitrogen, oxygen, carbon and hydrogen are the major substances needed to perpetuate life. Available in the air we breathe, these elements constitute almost our entire atmosphere. Ninety-six percent of the food consumed by plants and animals is built from these four basic elements, which leaves four percent for our consideration. This remaining four percent of food materials comes directly from the soil—and though it may seem a small portion, it is indeed a very essential one in sustaining life on this planet.

The soil minerals are as necessary to life perpetuation as are the four major elements. These minerals include phosphorus or phosphate, calcium or lime, sulphur, potassium or potash, and magnesium. In addition is our requirement for what are called the trace minerals, so named because they are needed in such small (but nonetheless important) quantities. Knowledge of these trace minerals is just now coming to life in its fullest measure; in fact, lack of awareness of their importance was one of the original factors in fallacies of chemical fertilizing. These elements are copper, zinc, iron, manganese and boron—none of which were attended to in von Liebig's experimentation.

There is some controversy about the difference between "organic" and "inorganic" minerals. Some scientists feel that minerals by their very nature are inorganic and, technically speaking, they are right. However, this kind of observation leaves out the results of certain processes on these minerals. Dissolved mineral compounds, initially formed without any plant or animal substances (generally having come from original parent rocks), can only be used for growth in solution. To allow this to occur, these minerals are processed within plants into new compounds which add the element of carbon to their make-up. The inclusion of this new element renders this originally inorganic substance organic. However, when the processes of decay liberate these minerals into the soil, they are simplified again into purely inorganic elements. Such is the swing of nature, from simple to complex and back to simple again.

Research has shown that there is a constant cycle of mineral circulation between the upper and lower portions of the soil— that is, where there is soil that is strong and healthy. To keep the soil's workshop functioning at optimum level, this circulation must be thorough and complete. The best insurance that this will occur is the proper working of the fertility chain. Any soil which has had its fertility chain disrupted will begin to show signs of mineral deficiencies. At the same time, mineral deficiencies will be a distinct hazard to the other links on the fertility chain. The addition of isolated, artificial mineral fertilizers is not the way to enhance mineral quality and quantity, or to repair a lagging fertility chain. On the contrary, most chemical substances are so strong that they cause definite upsets in the natural mineral circulation. What is needed, when a mineral deficiency is apparent, is a good organic adjustment. By replenishing organic material in the soil (truly the cure-all in organic gardening) this adjustment can be accomplished and the entire fertility system eventually restored to a good functioning level.

It is the *quality* of the minerals in the soil that is really vital in perpetuating a strong fertility chain; the quantity is less important. And nature, if left to her own devices, will provide the soil with the highest quality of minerals available. In fact, for normal purposes, nature provides sufficient amounts of everything necessary to keep the fertility chain going. It is man's mismanagement of the land and artificial methods that have drained nature of her own capacity to meet any reasonable demand.

Minerals naturally present in the soil will act at a pace in accordance with the natural balance and requisites of the fertility chain. Artificial mineral substitutes, however, are not at all tuned in to the tempo and required rhythms of nature. Such fertilizers are too readily soluble—some of these compounds go into spasmodic action upon contact with the soil. This is a sure way of throwing the soil workshop activity into utter chaos. Our chapter on chemical fertilization speaks more directly to this problem.

Only the activity of bacteria at work decomposing organic matter can provide soil with the entire host of major and minor mineral elements it requires. Without any one factor in this company of substances working together, entire mineral activity will falter. And when the mineral activity is not up to par, the fertility chain begins to suffer.

IN SUMMARY

The fertility chain is a rather complex and highly dynamic phenomenon. Composed of the interworkings of soil microbes, fungi, algae, earthworms and minerals, it is an intricate pattern and a highly efficient one. Efficient, that is, if it is permitted to follow its own natural laws; and even more efficient if its workings are enhanced through the methods of organic farming. It is impossible to maintain an unbroken fertility chain if adverse factors in the soil dominate the situation. Any powerful disruption of nature's cycles will seriously hamper the constructive natural processes carried on by the fertility chain. An emphasis on any one aspect of the processes or a disregard for a link on the chain will throw the entire system out of balance. It is this insensitivity to the necessity of balance and unity that has comprised man's major mistake in chemical and artificial farming. If a suitable environment is not provided for each and every member of the fertility cycle, they will lose potency and render the rest of the chain practically impotent. Nature provides automatically for this proper environment and achieves the perfect balance without a second thought! But the thoughts of man have not included the cyclical workings involved, and have thrown the processes seriously out of kilter.

Fertile soil is naturally balanced soil, and naturally balanced soil requires careful maintenance of the fiberization of the spongy vegetation layer of uppermost soil. If this sponge is not in evidence, it is certain that the entire dynamics of the fertility chain will suffer due to discordance in the soil world. Chemical interference by man spells sure disruption of the spongy surface of the soil, as well as trouble for all the vital aspects of the fertility chain.

ACIDITY AND ALKALINITY

Whether a soil is more predominantly acid or alkaline depends on the kind of rock from which it has been formed, as well as the partial or complete decomposition state of its vegetation. Soil productivity is very much interrelated with the acid or alkaline predisposition of the soil. This is expressed in what is called the pH factor, the technical name for the relative degree of hydrogen in the soil. A soil which is extremely strong in one direction is bound to have a poor effect upon soil fertility and subsequent vegetable growth. Soil that is heavily acid or extremely alkaline has an injurious effect on almost all crops.

In a soil too strongly acid, bacteria necessary for plant tissue decomposition cannot function properly. From our understanding of the necessity of the decay process in vital plant growth, we can readily understand why this is a liability. Furthermore, highly acid soil makes the elements manganese and aluminum so soluble that they are absorbed too quickly by the plants and throw the growth process way off balance. It is interesting to note that one of the basic properties of chemical fertilizers is that they are combined with acid in order to make their ingredients quickly soluble. Because the process is accelerated beyond the proper natural scope it brings chaos to the entire growth cycle.

A soil that is too heavy on the side of alkalinity will render plant nutrients almost unavailable. If too strongly alkaline, a soil will begin to lose its structure and will disturb the process of growth. It used to be that farmers were so in tune with the character of their land that they had only to taste the soil to determine if is pH factor was conducive to fertility. Of course, back in those days, it was not known as a pH factor—a farmer knew that if his soil had a sour taste its acidity level was too high. If he tasted a bitter flavor he knew that the soil was too alkaline. The suitable condition of the soil was signaled by a sweet taste—a firm indication that his land was prepared to bring forth a rich and healthy crop. Unfortunately there are few such farmers remaining today who are able to determine the condition of their soil by taste. There are several ways for modern gardeners and farmers to determine the pH factor in their soil. For more ambitious gardening, it would be well to purchase a soil test kit, available on the market at a reasonable price. Another alternative is to send a sample of your soil to a laboratory designed to test it.

Gardeners who seek a quicker, easier and more interesting method of soil testing can purchase litmus paper—easily obtainable at most drug stores. If you touch the paper to your soil you can determine on which side of the acid/alkaline balance it is by the color the paper turns. If of an acid nature, the soil will turn the paper pink; if too alkaline the paper will turn blue. A neutral soil condition (in which the acid and alkaline composition is just about equal) will be indicated if the paper remains its original color.

There are several remedies available for the organic gardener seeking to rectify the acid/alkaline situation if he finds it unsatisfactory. By working within the master plan of nature which is constantly striving to maintain proper balance among the living things on earth, we can help create an equilibrium conducive to the growing situation.

To neutralize a soil too acid in content, it is necessary to make sure a good amount of nitrates and other plant nutrients are available. This can be accomplished by making certain that the microorganisms within the soil increase their activity of converting nitrogen into nitrates, thus stepping up the decomposition process. This in turn can be brought about by a healthy application of organic matter to the soil. Another excellent way of correcting a too-acid soil is to mix crushed limestone, wood ashes or even ground oyster shells into the compost heap. Lime will work naturally to restore a more favorable balance to an overly acid plot of land and if it is applied within the compost (and our explanation of the construction of compost points this out as essential) it will reach the land at beneficial intervals and in its best working state.

Over-alkalinity can be rectified by adding organic matter to the soil also. It might seem strange that the correction for both conditions lies in the same remedy, but if we look at the situation more closely we will see that it is only natural! The process of decomposition releases acids into the soil and the presence of excess alkalinity will be neutralized by action between these acids and the soil. If provided with enough organic matter to remedy an off-balanced situation, nature will ultimately make the correct choice. Natural selection is a phenomenon man cannot duplicate.

So regardless of the state of the acid/alkaline condition of your soil, using plenty of good, organic material will serve as successful first-aid. It is good to know, however, that some vegetables thrive better in a slightly acid soil and others are more productive if grown in a soil slightly alkaline. If rectifying the situation of imbalance (as long as it is not too severe) seems impractical, a gardener can decide to grow acid-favoring vegetables if his soil is acid, and alkaline-loving vegetables if his land leans in that direction. To simplify this decision-making, we've listed the preferences of several vegetables.

Plants which favor a slightly acid soil: asparagus, beets, cauliflower, celery, onion, rhubarb and squash. Plants which will appreciate a slightly alkaline soil are: corn, tomatoes, potatoes, cucumber, lettuce, cabbage, beans, peas and most domestic garden crops.

If you have reason to believe that your soil is not strongly imbalanced in either direction, you might go ahead and plant what you like. It is helpful, however, to procure this information for the most successful gardening results.

SOIL DEFICIENCIES

There are several situations which lead to the various kinds of deficiencies soil can experience. One of the most common causes is erosion, a condition which results from an imbalance in the processes of nature that work in conjunction with the soil. Mismanagement of land and general ignorance of the need to assist natural processes have been in large part responsible for the escalating erosion situation in this country.

Erosion can come about when the protective covering of plant life (which we have referred to as the spongy surface) is removed from a section of land. As a result, this area is exposed to the ill effects of too much water and wind. Because it is in a high state of vulnerability, action on it by these elements will cause it to erode. We can see the process of erosion in forests which have been cut down with no provision for a restoration plan. The trees and vegetation which kept the soil surface intact have been removed so that the soil is left naked and open to extremes of weather.

Any disturbance of the natural vegetative covering of soil will ultimately lead to erosion. This can occur for various reasons, but is basically due to a disruption of the fertility chain or to artificial additives which inhibit and often poison surface plant growth. We cannot emphasize too much how crucial maintenance of the spongy topsoil layer is; in every aspect of good gardening we are at some point dependent on the good auspices of this portion of earth.

Erosion can also occur when soil becomes hardened (this is usually a situation that appears concurrently with a loss of vegetative sponge) and it can no longer absorb and pass off water. When soil structure is loose and fibrous it can readily absorb and make use of the water it comes in contact with. In the case of hardened, tight-packed soil, the water accumulates on the surface and then washes away all at once, taking with it a good deal of the precious topsoil. This hardened condition often results from the continued application of chemical fertilizers which simultaneously destroy vegetative mulch covering and compressing the soil.

The process of erosion is a deadly one—whether it is occurring in the huge forests of the earth or in your own tiny garden plot. The build-up of topsoil is so slow and tedious that to wash it away so carelessly is liable to mean the non-productivity of land for years and years to come.

Another major cause of erosion is a lack of organic matter in the soil. This is connected to the "hardening up" that caused stream (water) erosion, and is again a result of the loss of spongy protection. Hardened soil "locks up" vital nutrients, rendering them unassimilable to plant life. To avoid this hardening of the soil, and the loss of the spongy covering, it is necessary to replenish the organic content of the soil through organic methods. In fact, any soil ailment, any negative condition you may find in the earth you work with, can be successfully treated in the same, simple, natural and healthfully basic way—by observing the law of return—giving back to the soil that which has been taken away. Often man takes more from the soil than he imagines, simply by growing and harvesting vegetables year after year. To ignore the need for return, and compound the error with the application of chemicals is the most severe mistake we can make where the cycles of life are concerned.

Another problem often evident in depleted land is specific mineral deficiencies within the soil. These ailments, too, can render your crop lifeless and without vitality.

A phosphorus deficiency retards actual plant growth. It will take longer than it should for phosphorus-lacking plants to produce a crop. Such a deficiency is indicated by a reddening of the soil.

A calcium deficiency results in yellowish plants with the stems exhibiting a thick, fibrous material. The best treatment for a calcium deficiency is to coat the ground with limestone.

Any sort of deficiency will respond excellently to the addition of organic materials. We can see that the buoying up of the soil through conscientious enhancement of natural processes is, in countless ways, a safeguard as well as a fertilizer. Naturally resistant, sturdy soil will in turn produce strong and resistant plants, thus lessening the fear of disease or insect devastation. These facts are not conjecture—they have been carefully documented by scientific research and observation!

IN SUMMARY

Understanding the life functions within the soil will assist you not only in your theoretical understanding of gardening, but in your everyday dealings with the soil. When we grasp the ever-winding, constant flow of the fertility chain and other processes within the soil, it becomes evident that organic gardening is the only answer to preserving our environment.

Microbes, fungi, algae, earthworms. soil minerals—all these

play an important role in the continued health and wealth of the soil—they must be allowed to function naturally and without interruption. The addition of organic material will not only encourage the smooth workings of the vital processes—it will go far in assisting and strengthening them.

To understand completely the characteristics of your garden soil, you might want to test its pH factor as we have discussed— either with soil testing kits or litmus paper.

When you begin to feel the tremendous interdependence of all living things and begin to realize exactly what transpires in the earth to promote growth—you will be well on your way to becoming a gardener with soul!

For more information on the make-up of soil and the workings of the fertility chain, refer to the following: ORGANIC GARDENING & FARMING, Joseph A. Cocannouer, *Arc*; INTRODUCTION TO ORGANIC GARDENING, Chuck Pendergast, *Nash*; FARMING WITH NATURE, Joseph A. Cocannouer, *University of Oklahoma Press*; INFLUENCE OF SOIL MINERALS ON INSECTS, Leonard Haseman, *Journal of Econom, Entomol.*, #53; PHYSICAL, CHEMICAL AND BIOCHEMICAL CHANGES IN THE SOIL COMMUNITY, William A. Albrecht, *University of Chicago Press*; MAN'S ROLE IN CHANGING THE FACE OF THE EARTH, Wenner-Green Foundation for Anthropological Research, *University of Chicago Press*; THE BASIC BOOK OF ORGANIC GARDENING, Robert Rodale, *Ballantine*.

OUTDOOR GARDENING

The homeowner with a backyard (or front yard!) large enough to provide some gardening space is indeed a fortunate person. Don't make the mistake of thinking that you haven't enough room simply because your yard isn't exactly spacious. It's good to keep in mind that just about any amount of space can be a potential garden, if you approach the idea with energy and optimism. Consider yourself blessed to have even four square feet of our good earth to turn your hand to—and remember that many of your comrades in this business of organic gardening are doing quite well with no outdoor space at all! Observing the processes of nature and learning the basic principles of her activities will turn you on to the flow of life *regardless* of how little space you have to devote to your gardening project. It's quality rather than quantity that we're aiming for here—quality and the kind of multi-leveled understanding of nature's ins and outs so many of today's gardeners and farmers are lacking. We're going to lay out general instructions for garden plotting and count on you to figure the necessary adjustments to your own capacity for allowable space. And don't forget that being an outdoor gardener doesn't preclude your dabbling in the indoor art of gardening. In fact, your work outdoors will give you the experience you'll need to carry on year-round gardening indoors once the frost has set in and you lay your outdoor plot to rest for the season.

When contemplating an outdoor garden, it's a generally good rule to think in terms of early planting whenever possible. Early planting will mean, in some cases, that you'll be able to harvest a double crop—that is, you can gather an early crop and plant a second in its place. This is what is called successive gardening, and we have gone into specific modes of working this method in our chapter "What to Plant." There are certain combinations that work well together in successive planting, and you'll want to keep them in mind because they often have direct bearing on the health of your soil.

Early planting and subsequent double cropping are special

assets in small gardens where space is necessarily limited. Never leave any space unplanted if you've harvested in mid-summer— if you put your mind to it, you'll think of a vegetable that will grow quite nicely in the spot an earlier crop has vacated. You only need to be sure that enough time is left for the crop to mature properly. In some cases, this won't be much of a worry because you'll be able to plant cold-hardy crops that can stay on the ground either through the winter, or right up until the first heavy frost. Once you understand the characteristics of each vegetable in terms of its temperature and weather tolerance, you'll be able to plan successive planting with confidence and excellent results.

Another aspect of early planting is that it takes into consideration the very vital latent growing energy present in your soil. Ground that has lain dormant all winter is just aching to bring forth fruit and it's a good idea to take advantage of this energy. Consider it a gift from nature and don't let it go unused—it will serve you well.

There are many vegetables which are so hardy to cold that they can actually be planted a month or so before the average date of the last freeze, or six weeks before the first frost-free date is estimated. Such cold-tolerant crops often require protection from heat rather than cold, so getting them in the ground early also serves the purpose of avoiding a confrontation with mid-summer heat. The point being that these crops will be harvested prior to the hottest part of summer, and, by employing the successive gardening principle, heat-loving vegetables will have been planted in their place. Understanding these cycles is just one more way of aligning yourself with nature's laws and at the same time getting the most out of a very willing earth! Cold-tolerant plants should not be planted in late spring, but in very early spring, or in late summer, when they will grow to maturity during fall's cooler temperature.

There are a good number of popular vegetables which have this quality of cold tolerance. They are: broccoli, cabbage, lettuce, onions, peas, potatoes, spinach and turnips. You'll want to get these in the ground early and prepare to replace them with some of the heat-loving plants. Other vegetables which are also quite hardy, but not to be planted more than four weeks prior to the first frost-free date are: carrots, beets, chard, mustard, parsnips and radishes. These plants should not be left in the ground too late into the fall as they are not quite as frost resistant as the earlier named types.

There are vegetables which you should take care not to plant before the first frost-free date. These specimens will not stand up to the ravages of cold weather. They are: snap beans, cucumbers, okra, spinach, soybeans, squash, sweet corn and tomatoes. Once the danger of frost is gone you can plant these without a worry and expect them to grow strong and healthy.

Still other plants require hot weather to grow properly, aside from their general inability to withstand cold temperatures. These should not be planted until at least one week after the frost-free date. Included in this category are: lima beans, eggplant, peppers and sweet potatoes. These vegetables can be effectively planted in the late spring or during the summer. Planting itself is a cycle— as the peak of summer heat passes, you can once again think about planting the medium heat-tolerant varieties, and towards summer's conclusion, go back to the planting of cold-hardy crops. Usually you need allow no more than six to eight weeks before the first fall freeze for these weather-tolerant vegetables to mature. By understanding the relationship of vegetable-growing to the vicissitudes of weather, we can plan for the largest and healthiest crop possible. It's one more facet in the very broad scheme of the whole scope of nature. We must set ourselves to understand all of the various circumstances that need consideration for the most successful garden possible. To overlook even one of these is to render the opportunities for good gardening less effectual. At the very basis of organic gardening principles is this idea of totality—of regarding the entire process of growth as a universe unto itself— very whole, very complete and very structured in its own simple way.

You might want to start some of your plants indoors before setting them out in the garden. Towards this end, you can experiment with hot beds and cold frames. These are structures employed to provide for seed germination prior to outdoor planting. In this way you are setting young seedlings into the ground, rather than seeds themselves.

It used to be that almost all gardeners and farmers employed this method of growing their own plants rather than buying young seedlings from a gardening supply store. Nowadays, people who prefer to work with plants rather than seeds and don't want to germinate the plants indoors, are buying commercially grown plants. Most organic gardeners agree that this is not a good idea because you just can't be sure how these seedlings were germinated. You can't be certain what processes were employed in the nurturing of the seeds into young plants.

Germination is the process of providing a seed with a perfect total environment so that it may grow into an independent plant. Until a seed has a root and leaves that are green, it is not capable of making its own food. The germination process takes from 10 days to a month, at which time the plant can produce food in its own structure. A seed is actually an embryo plant with enough food stored within it to see it through the germination process. When a seed possesses its own root that root can take up waste and minerals from the soil. At this point these substances are in the form of raw materials, and are carried by the plant's sap into the leaves where they are made into plant food. It is necessary that these leaves be green because only chlorophyll, the substance which causes the leaves' "greenness," has the ability to transform the raw materials (minerals and water) into starch in the presence of light. Both chlorophyll and light are necessary factors in effecting this transformation. It is in this way that the leaves and other green parts of the plant act as a kind of factory for the preparation of plant food. Once this process begins to function the plant is an independent entity. But until the plant has reached a point where it is capable of this self-sufficiency, the cold frame or hot bed will provide the necessary controlled environment. This will include providing the germinating seeds with proper warmth, light, food and moisture so that they can grow strong and mature enough to attain this independent status. Seed germination is one of the most exciting and delicate aspects of farming, and it's an excellent way of giving your vegetables a head start when it's too cold to plant outdoors. When the time comes and you are ready to transplant the vegetables to their outdoor, permanent home, they will be sturdy and independent. And of course, you still have the option of letting seeds germinate in the ground itself. In our explanation of specific planting procedures, we've allowed for this choice. You'll find directions for planting both seeds and young plants as the instructions for each vary slightly.

Hot beds and cold frames are basically identical structures with the same idea in mind—the care and nurturing of seeds during germination. The difference between them, obvious in their names, is that hot beds serve as homes for germinating heat-loving plants such as peppers, tomatoes and eggplants. Since you can't set these vegetables out, even in seed form, until the weather is rather warm, starting them off in hot beds will save you time. Cold frames are constructed exactly like hot beds except that they can be kept outdoors unheated—protected by either a covering of boards or a glass protection. You can employ a cold frame for use during the first days of frost in the fall, and thus provide yourself with fresh lettuce and other salad greens that are cold-hardy. You can use this handy device right up until winter. And many gardeners use their cold frames in early spring, when they're not certain that the frost is gone for good. At either end of the growing season, cold frames are tremendously useful and will expand your harvest, if used properly.

Another use for cold frames is as places to store vegetables. You can accomplish this by removing about 18 inches of soil and laying a layer of straw in this pit. On the top of this straw place vegetables such as turnips, rutabagas, beets, carrots or celery. Then put another layer of straw across the vegetables and cover the whole pit with a row of light boards. This structure will keep vegetables crisp and unfrozen all winter and is an especial asset if you're lucky enough to harvest a crop that is in excess of your immediate needs. In subsequent pages, you'll be learning which vegetables lend themselves best to in-ground storage and how to go about this most successfully.

In speaking of the process of germination it is well to carry our discussion one step further with some advice on the transplantation process that follows successful germination. Once you've got your seedlings to a point where they're exhibiting their own roots and green leaves, and the weather outside is right, you'll want to start getting them into their permanent outdoor home. Of course, if you're gardening indoors you can germinate your seeds right in the container or trough that they'll be staying in permanently, since the transition to outdoors won't have to be made.

The initial and most important procedure in transplantation is what is known as "hardening off." This is a way of gradually acclimating your seedlings to the natural elements they're soon to be exposed to full time. If you simply move a plant outdoors from an indoor seed bed, the sudden shock of temperature and light would kill it almost instantly. The best way to accomplish hardening-off is to move the plants outside right in their flats when the weather stars to warm up. At this point keep a protective covering over them. Leave it on completely for at least two days when the beds are first outside. Then gradually begin to expose the young plants to the elements. This is initiated by opening the covering ever so slightly on the third day the beds are outside. Each day open the covering even further for the next two weeks. At the end of this two-week period the plants should be totally exposed and are then ready to be placed in the ground. At that time they will have become used to the weather in a way that would have been impossible without this gradual introduction.

The major difficulty we're dealing with in transplanting is the element of shock. Even if you've successfully hardened-off, moving a plant from one home to another can be a tremendous jolt to its entire system. Each plant is a microcosmic example of the entire universe of living things—that is, each plant is a miniature universe all by itself. Plants function in a highly integrated fashion, every cell being part of an interdependent chain of other cells and cell activities. We witness similar interdependence in the function of the fertility chain, and indeed in all of nature's processes. Transplanting is a disturbance of the plant's smooth-running universe. When a plant is lifted from its place, vital roots and tiny root hairs are unavoidably torn away and this disrupts the entire flow of water and soil through the roots, stems and leaves. Aside from this interruption of plant and cell nourishment, the sudden change in light and other vital conditions will be a great shock to the entire organism.

It is up to the conscientious gardener to provide for the easiest adjustment possible subsequent to transplanting. It is best to go about this procedure on a cloudy, moist day when the light and heat shock will be minimal.

Begin the process by loosening the soil in the new location and working it over properly. What you do at this point will depend on the work you think is necessary in terms of fertilization and breaking up of the earth. Perhaps you have seen to this earlier, and if you have done so, you are certainly a wise and conscientious soul! Also loosen the soil around the plants in their beds, watering it prodigiously about two hours before removal, to make it as easy as possible.

When you do finally lift the plants out of the soil, take as much of the surrounding dirt as possible, creating a minimum of disturbance to the roots. Unfortunately, even the most careful transplanting is likely to damage some of the roots, but caution will help keep this destruction to a minimum. When the plants are placed in their new location, make certain that the roots are in firm contact with the soil, in order to assure maximum absorption of what the ground has to offer in the way of nutrition, water and other vital substances. Keep an eye on your seedlings after transplant— they should be well watered and shaded at least for the first few days, and especially if the weather is uncommonly hot and sunny. In this case provisions for shade should be made for the plant's first week in the soil. Remember that it's going to take awhile for the plants to become accustomed to their new home and everything you can do to lessen the shock and make them more comfortable will be towards the interest of a healthy and sturdy crop.

Plants which lend themselves well to transplanting are broccoli, cauliflower, lettuce, cabbage and onions. With plants like beans, eggplant, peppers, tomatoes and celery, you're liable to encounter a greater sensitivity to the change and you might lose a few in the process. Plants with taproots are most difficult to transplant and it is quite a feat to reestablish the proper root-soil relationship. These plants include carrots, cucumbers, corn, melons, squash and beets.

LAYING OUT THE PLOT AND SEEDING IT

Before we get into the dynamics of laying out your plot and seeding it, a reminder about compost is in order. If you've attempted the Indore compost heap, or even a variation on Sir Albert's theme, then figure on having it ready about a month before you're going to plant. For general soil enrichment this is the ideal time for applying the rich, life-giving humus you've worked so hard to create. If you can't get your compost on the land this far ahead of time, just make certain that it's well ground up and shredded when you do lay it down. This will increase its availability to the soil by assisting in the breaking-down process. It's a general rule that the more refined and worked-over organic matter is, the quicker and more easily it will lend itself to the soil-enhancing process. There are some garden cultivating tools which will be of great assistance to you in this task—some will help in shredding the compost, and others will be ideal for aiding the spreading and mixing process.

Compost should be thoroughly mixed with the top four inches of soil, at least. For plants that send roots down further, you'll want to work the compost in deeper to allow it to nourish the root systems below the surface of the ground. It's almost always a good idea to apply this already decayed material as heavily as you can, since having decomposed prior to its addition to the soil, it won't drain the land of necessary nutrients but will only give strength and vitality. You can gauge the amount you'll want to use by assessing the fertility level of your soil as closely as you can. Naturally, soil that is depleted and lifeless will need a bigger boost from compost material and a more fertile soil will be able to do with considerably less. One of the beauties of the organic method is that the benefits are cumulative—that is, as the years pass you'll see your soil becoming more and more healthy. Chemical farming has just the opposite effect on the land—as the years pass the soil becomes more sickly and increasingly less able to support plant life. In any case, this addition of organic matter to the soil is basic to proper soil preparation and good planting procedures.

When you've decided on the shape and dimensions of your garden area, mark the corners of the plot with some kind of permanent markers—stakes or pipes will serve this purpose well. Set up a first line, marking where it will fall with a rake handle or your hoe. It's a good idea to plant in straight rows—not only for appearance's sake but for general organization of the mulching and weeding procedures. You'll probably wind up with the least waste of precious space if you attempt to keep the rows fairly straight and parallel to each other. To insure that the rows are as straight as possible, set up a stake at the end of a proposed row and stretch a string above the soil's surface to the other end. When it looks straight, place the other stake and remove the string.

Once you've got the row marked, you can use the shallow furrow you made with the rake handle for planting small seeds, and increase its depth for larger seed plantings. Most vegetable seed should not be planted too deeply, in fact seeds for lettuce and radishes will not be put into the ground deeper than one-quarter inch! In most cases you won't need more than a depth of one inch, even for larger seeds, such as peas or beans. We've given exact planting depths on all vegetables in our chapter on choosing vegetables. As a rule late-planted varieties will go into the ground more deeply than early types. This has to do with the receptivity and looseness of the soil which has been increased as the weather gets warmer. Soil that tends towards heaviness requires a more shallow planting of seeds while lighter soil can receive seeds beneficially at a greater depth. When temperatures are low, it's a good idea to plant seeds shallowly, increasing depth of planting as the weather gets warmer.

The actual dropping of the seeds into the earth—the sowing—is one of the most significant and moving parts of the entire gardening process. It is difficult to express in words why this is so, perhaps it is the joy of feeling yourself a part of the very birth of something—sensing yourself an aid to nature's processes. It's a highly beautiful and symbolic act that most people find extremely gratifying. Part of the joy, too, is the knowledge that your time and efforts will be rewarded by the production of food that you have seen through from the very beginning. Whatever it might mean to you, take some time with this beginning step—relax and feel yourself a part of the enormity and simplicity of nature—it's an experience you'll not soon forget, we can promise you that!

In sowing the seed it is well to keep moderation in mind in most cases. Seed should be spread thickly enough to get an even growth, but sowing seed too thickly will only result in waste and extra work in thinning. Once the seed is spread (and you can accomplish this by either shaking seeds from a ripped corner of an envelope or dropping them in one by one) draw the soil over the seed with the corner of the hoe, or you can smooth it over with your hands. Many times you'll be instructed to cover the seed with compost that is measured out for this purpose. If the soil (or compost) seems cloddy, it's a good idea to break it up a bit before covering the seed; this is a task you can accomplish ahead of time if you observe that it will be necessary.

If the soil is heavy, be careful not to pat or pack it down too much—this is likely to lead to crusting of the topsoil which will not lend itself well to water and air circulation and will ultimately jeopardize the crop. Tightly packed soil is never an asset—but firmness and tight packing are not necessarily synonymous conditions—use your intuition to guide you here. Sandy soils may require a bit more firming after planting, so keep the soil type in mind while planting and take it from there.

There are three basic methods for planting seeds. The *drill method* is the best and most widely used of these. The corner of the hoe is used to make a straight drill (trench) down each row. This is made deep enough to make contact with moist ground. The depth will vary with each vegetable that you plant, too. Remember that the ideal distance between plants in a row refers to the result after thinning—you won't want to plant your seeds two and three feet apart—you'll just want the plants to stand that way once they really begin to flourish. Specific distances are listed in the vegetable section.

A second method of planting is the hill method—usually employed for plants that are going to spread out on the ground. If you're working with a small garden, or indoors, it's not likely that you'll be involving yourself much with these types because they are great space users. If you'd like to experiment with the hill method, however, dig a hole about a foot wide and a foot deep. Then lay a six-inch pile of well-rotted manure or mature compost at the bottom. Next lay in six inches of fine soil and plant the seeds at the top of this layer. Again, the depth of the seeds will depend on specific instructions. Another advantage of the hill method is that you're supplying a good deal of extra nutrients by digging the six inches of manure below the plants. In basic gardening lingo, by the way, "hill" refers to any raised area that contains three or more seeds after planting. If you're fortunate enough to get all the seeds in a hill to grow, you'll want to thin the plants out once they get started.

A third method of seed sowing is the broadcasting method. This involves scattering the seeds thinly over the top of the soil and then lightly raking soil over them. This is usually employed with very small seeds which should be planted at shallow depth and not covered over too deeply. Turnips, lettuce and similar seeds will lend themselves to broadcast planting, if you'd like to give it a go. Remember, though, that seeds sown in this way will not produce the straight rows that trench or hill planting will. This is something to keep in mind if you've a fancy for organization.

It might be a good idea to try each type of planting, just to see which gives you the best results or which you find most enjoyable. In all phases of gardening, experimentation is an excellent plan and will provide you with valuable learning experience. If you've got the time you might attempt to keep a journal listing your planting procedures and how they worked out for you. You could also make notes on anything you learn through the trial and error method or any pointers you want to remind yourself of the next time you plant. This journal can be very comprehensive—including all aspects of your gardening experience—procedures, choices of vegetables, dates, composting and harvesting hints. It will also provide you with some interesting reading in years to come! You might also want to keep a record of the varieties of each vegetable that you find especially hardy or tasty—it might be difficult to remember these specific names otherwise. Such a journal will insure that the trial and error method need be employed only once on each question you have.

THINNING

Once you've planted the seeds and your young plants begin to raise themselves above the earth's surface, you'll want to think about doing some thinning. Especially where an overabundance of seed has been used in initial planting, the thinning process will be a necessary and important one. You can save yourself a lot of time and energy by trying to be moderate in your sowing—but don't underdo it either—too thin a spread of seeds is liable to result in a scanty stand. Your own good judgment, and of course an accumulation of experience as you continue to garden, will stand you in good stead on this subject.

Thinning is most easily accomplished when the plants are still small and also at a time when the ground is moist—this will enable you to pull out the extra plants easily without injuring the ones that are to remain. A struggle in thinning or weeding is likely to harm the root systems of plants you are anxious to protect. You might also be eager to preserve the root structures of the plants you're pulling out, if you're considering using them to begin a new row of plants somewhere else in your garden. And even if you're going to use the thinnings in the kitchen (which for many vegetables is a distinct and delicious possibility) you'll want to have them in the best shape possible. It's a good rule to employ gentleness when dealing with anything that grows—so take your time, and relax—violence certainly has no place in your serene and peaceful garden plot!

If you've got a crop of taproot vegetables, it's a good idea to thin them before their roots become fleshy. Such root development will make thinning a difficult job by the time you get around to it; you may, however, permit onions and radishes to stay in the ground till the thinnings are large enough for kitchen use. This will provide you with some early produce and also cut down considerably on waste. But actually there need be no question of waste in the organic garden—thinnings that you won't be replanting or using in the kitchen can be added to a pile of green matter to be composted. In the organic garden there's a place for every living thing—whether or not it seems superfluous to you. The cycle of life allows for all organic substances to at least return their life-giving elements to the soil for re-use by new growing things.

You'll be mostly coping with small-seeded vegetables during the thinning process. These are the types that are likely to have been overly seeded and require some thinning out. Swiss chard, for instance, can be thinned at a height of about three inches—and then as the plants develop, plan to harvest some a bit early (you can alternate plants) to provide more space for the ones you'll allow to mature. Melons and cucumbers should be thinned to stand about one foot apart, or if planted in hills, should be thinned to two or three plants per hill for best results. If the thinning process seems to be a bit of a shame to you, just keep in mind that only if plants are permitted enough space without excessive competition for soil nutrients, will they grow to their greatest stature of healthiness and hardiness. Be firm in your decision to allow each plant its necessary space allotment—in the end, you'll be glad you did!

Carrots can also be thinned at about three inches, so as to stand about one inch apart. When they are large enough to eat, alternate plants can be pulled to provide more space for the still-growing carrots.

Beans, peas, corn and other large-seeded plants don't usually require thinning. You'll be able to judge the number of seeds you're planting quite easily, because these seeds are easily counted and divided accordingly. You need only plant a few seeds more than the crop you're hoping for. At no time in the sowing process is "the more the merrier" the way to think—it's liable to be just so much wasted seed and space if you overdo it.

And now—our soil has been prepared, our compost has been spread out and worked in, the crop rows are marked as we've decided. We've furrowed into the earth and dropped our seeds in, or brought young plants out to the garden from their indoor flats. We've patted the soil over the seeds and even stepped back to admire the fine, straight rows we've toiled over. The basic work has been completed and the projects we've got ahead in the garden will be mostly a matter of employing the work of thinning, weeding, mulching and perhaps a bit of compost side-dressing. We'll also have to beware of those exceptionally dry and sunny spells—providing adequate water for the moisture-loving plants when such weather hits. If we've planned to work with the concept of successive planting, then there will of course be some mid-season harvesting and re-planting according to the best rules of nature and the cyclical growth of different plants.

One of the most exciting and rewarding parts of gardening is awaiting the first tiny bit of evidence that something is going to grow from that small seed we put into the ground! So watch your garden carefully—not in a worrisome or impatient way, certainly —but with the wonder and interest that the whole process deserves. You might want to add notes to your journal as you observe the growth of your garden—perhaps there are things you see you should have approached differently—perhaps a hunch of yours will glean rewarding results. Stop, look and listen—tune in to the workings of nature. After all, your garden wasn't a project you undertook simply for the exercise of bending and raking—there's some soul-soothing vibrations for you to pick up on and you deserve the many pleasant moments they'll bring you. A stroll in the garden, anyone? Even if your plot is small, you'll find yourself showing it off to your friends and neighbors, and as time passes and your enthusiasm grows—you'll be eager to help them do some experimenting on their own!

We've given you the basics of laying out your home-garden outdoors. In the next chapter we've worked to assist the indoor gardener in adapting this information to his own gardening work. So, if you're interested in moving your gardening efforts and your rapidly growing green thumb expertise indoors for the winter, read on!

For some helpful and absorbing reading on the mechanics of outdoor gardening, we'd like to recommend: GARDENING WITHOUT POISONS, Beatrice Trum Hunter, *Berkeley Medallion;* HOW TO HAVE A GREEN THUMB WITHOUT AN ACHING BACK, Ruth Stout, *Exposition;* GARDENING WITHOUT WORK FOR THE AGING, THE BUSY AND THE INDOLENT, Ruth Stout, *Devin-Adair;* THE BASIC BOOK OF ORGANIC GARDENING, Robert Rodale, *Ballantine;* ENCYCLOPEDIA OF ORGANIC GARDENING, J. I. Rodale & staff, *Rodale Books;* HOW TO GROW VEGETABLES AND FRUITS BY THE ORGANIC METHOD, J. I. Rodale, *Rodale Books.*

INDOOR GARDENING

If you're an apartment dweller, with very little space to devote to an organic garden, perhaps you're feeling somewhat awed by our fairly in-depth discussion of the nature of gardening and the more ambitious methods of performing basic organic procedures. But we stated early on that this was a book that would start *everyone* gardening—that would ultimately be as helpful to the city apartment dweller as it would to the reader with a good amount of outdoor gardening area. Well, we promised, and we mean to keep our word! In fact, as we mentioned earlier, we're devoted to the proposition that gardening can be an experience available to everyone, at least to some degree. And if you've read this far and we've got you interested and eager, then no doubt you're anxious to learn how you can apply all (or almost all) of the previous information to your own humble organic gardening project.

The detailed explanations of the concepts of organic gardening and the dangers of the chemical approach that we've outlined have given you an idea of the basic set of principles and functions all gardeners work with. For no matter how tiny or insignificant *you* may think your indoor gardening experiment is destined to be, we beg to differ! The cycles of nature, beautiful and simply woven (though they may seem somewhat complex to you initially), are active in *every* living thing—regardless of its size or where it is grown. That is, every single vital natural process we've delineated—microbial action, nitrogen cycle, decomposition and decay—is going to be taking place right in your living room once you've got something—anything—growing there. This is your chance to bring some of nature's wonder into your own home—and it can be more than a venture in greening your apartment decor. You're going to wind up with some tasty, healthy tidbits for eating and an enhanced consciousness of nature's marvelous life processes. You're on your way to tuning in to the very core of what it means to be alive, and by approaching your gardening experience with natural methods and understandings in mind, you'll be learning a hell of a lot in the process. So take heart and read on!

There is much you can learn from our explanation of plotting an outdoor garden. So if you've just barely skimmed that chapter because you figured it didn't apply, try going back over it more carefully a second time. Remember the concepts of nature and growth are basic and all-encompassing, and though you'll admittedly have to make some provisions and adjustments due to the fact that you're working inside, there's still a great deal of value in understanding outdoor gardening.

Now that you've got a firm grip on the essence of organic gardening, its definition, guidelines and basic thrust, it's time to get down to your indoor plot. Your first question might well be how you're going to get the soil to start your garden with. Looking out your apartment window on a landscape filled with nothing but concrete, skyscrapers, fire escapes and lots of humanity is liable to make you doubtful about attaining even this most elemental supply. Consider: have you a friend located in a more rural or even suburban area who might not be opposed to your filling up a bucket or two with some dirt from his backyard? Failing that, you might think about making a journey to a park or country area where you could easily avail yourself of a bit of soil. Don't attempt this unless you're certain it's legal! Obtaining dirt from the greater outdoors is preferable of course to obtaining the potting soil you can buy in a gardening supply store. Unfortunately, most of these commercially sold soils have been sterilized to discourage the formation of bacteria. By now you know that such bacterial deterrents will discourage not only harmful bacteria, but are bound to impede the work of the friendly and helpful ones that are necessary for good plant growth. Already we are applying the basics of organic gardening to your as yet unborn indoor garden, and, if you're feeling that you'd like to stay as close to the organic ethic as you possibly can, we feel it's best that you stay away from commercial soils. There is the possibility however that you could obtain non-sterilized, organically treated soil for your indoor project from a supply store that specializes in organic gardening supplies. In fact, it would be good to locate such a store for some other gardening needs that are bound to come up.

The amount of soil you're going to need will, of course, depend on the size of garden you've got in mind, and in some cases, on the type of vegetables you're planning to work with. As we'll discuss in our "WHAT TO PLANT" chapter, different plants require a different depth of soil, though for the most part you'll be working with fairly shallow requirements in an indoor project. Remember, as ambitious as you'd like to get now that the bug has bitten (if you'll excuse the expression!) you are limited by space and area. By the time you've decided on the container that's going to act as home for your garden plot, you'll be able to estimate pretty accurately how much soil you'll need to fill it to the proper depth.

Once having secured the soil for your garden, you should begin to consider how you're going to cope with the question of organic fertilizer. In reading the chapter on composting and understanding the extraordinary importance of humus and organic material in gardening, you've come to be aware of how badly you'll need these elements in your garden. The traditional method of composting, more specifically the Sir Albert Howard Indore method, is obviously a task oriented to the outdoors, so there's a matter of adjustment if you want to share in the benefits of compost but don't have an outdoor area in which to work.

The problem of composting indoors is an impossibility insofar as the Indore method is concerned. This is partly because of the specifications of size (usually five or six feet high and five to six feet wide) that are an important factor in the functioning of the classic Howard compost stack. Another factor which makes it literally impossible to create the Indore heap indoors is the necessity of air circulation within the stack. The manner in which decomposition takes place within this heap is partially based on constant and thorough ventilation as well as the circulation of water, either through rain or purposeful watering. In the Indore method, the final product humus is dependent on this continual aeration for its formation. Therefore composting of this sort cannot be accomplished in an air-tight container which doubles as your garbage pail. No matter how religiously you store up wastes in such a container, it will not produce humus in the classical sense of the word.

However, there are composting alternatives for the indoor gardener. There is another method of composting, often employed outdoors (and especially when there is a very large area to be compost covered). This type of composting is called anaerobic, because it does not require the amount of air circulation the Howard heap does. One type of anaerobic composting is called sheet composting and is a method by which organic wastes are applied directly to the soil and permitted to decay within it. Unlike the aged compost heap that has already gone through the process of decay previous to its addition to the soil, in sheet composting green matter or fresh manure is added right into the ground. We'd better mention, before going any further, that you needn't get nervous at the mention of manure; although it is the ideal compost ingredient, for indoor gardening there are several convenient substitutes to be employed. As we've already mentioned, in the place of manure you can add bone meal, scraps of kitchen meat (no fat, remember) and even some high-protein soy powder.

By utilizing the method of sheet composting, and adjusting it to his own needs, the indoor gardener can still have the valuable humus factor to treat his soil with. All kinds of kitchen wastes that would normally have been merely a messy annoyance to be disposed of, can be incorporated into compost within the soil of your indoor garden. So, right within your own home, you've begun a very personal and very real ecology movement! Think of it. In the true spirit of nature you have used organic matter to its fullest and then, rather than wasting the remains, have observed the law of return and put these valuable leftovers back into the growing process.

Just as with your decision about soil, the amount of organic material you'll want to save for compost will vary with the size of garden you've got. When you get a feeling for what things work best toward producing and enhancing top-rate plant growth, you'll want to save these for your garden and let the less valuable substances go.

The one drawback of sheet composting is that an anaerobic method of this sort should really be accomplished within the soil *previous to the planting of the crop.* This is because the process of decomposition which will be taking place within the soil will require the help of certain nutrients. Because of this, the decomposition process is likely to compete with growing plants for this nourishment if the composting is taking place at the same time as the crops are growing. This can be especially deleterious to young plants which are not really strong enough to compete successfully against the strong and ongoing process of decomposition. So, for any kind of reasonably sizeable sheet composting attempt, indoors or out, it is preferable to mix the organic materials into the soil and let them begin to compost prior to the planting of your crop. When sheet composting is done outdoors this is usually begun about two months before planting. This might be just too long for you to wait, so in adapting the sheet composting method to indoors, just allow as much time as you can for the decomposition process to get under way. It will draw more nutrients at its incipience, so once it's begun and well on its way, it will be much less of a drain on the soil. You can accomplish this pre-planting composting in your home by mixing the kitchen wastes you've gathered with the soil you plan to use and setting these aside out of the way. You can cover the mixture with plastic, for sanitary purposes, if you like, and just let it go to work for itself. Once you've provided the right ingredients and a reasonable environment, the process of decomposition will gather its own steam and do an admirable job all by itself. In fact, this is the case with all of nature, on all levels—the reason our earth is hurting today is due to interference with the natural scheme of things. Perfectly capable of maintaining her own forces and fertility, nature has faltered only because of the interference of man—in terms of harsh chemical soil additives and sprays and because of generally unviable land management procedures. You can give nature a chance to show you its efficiency right in your own home garden. Just remember to mix the organic wastes well with the soil so that decomposition can take place at optimum speed. If at all possible, it's an excellent idea to put the kitchen wastes through a sieve before mixing them with the soil; this will speed up the process of decay; small particles are more easily acted upon than larger hunks of matter. It's difficult to figure proportions exactly when you're dealing with uncomposting matter in the first stages, but the ideal ratio for humus (what you'll have when the composting's complete) is about one part humus to three parts of soil. Do your best at approximating that proportion when you mix in the organic materials with the soil.

This kind of indoor composting, while it is admirably within the spirit of organic concepts, is admittedly a rather slow procedure for the eager, ready-to-get-down-to-it indoor gardener. You might feel that waiting weeks just for your soil to be ready to plant in is more than you can handle, so here's another adjusted alternative to consider in terms of composting.

You can mix in the kitchen wastes or whatever organic matter you have on hand at planting time. Again, we must reiterate that the processes of decomposition that will be going on the minute these substances reach the soil will vie with the growing plants for nourishment. Because of this you have to be certain not to apply a large amount of organic materials all at one time. What you might do instead is to think about staggering the addition of organic materials as you go along. If you do decide to add a bit of organic material at a time, be careful not to disturb the roots of the plants when you're working the materials in. You want to have the compost pretty well turned within the first four inches of the topsoil (and it's not a bad idea to keep this area loosened up by working it ever so gently anyway) but you must be very, very cautious about disrupting the root structure. Replenishing the organic matter you've added to the soil is a good idea, no matter what type of compost method you've employed. After decomposition has taken place and the simplified life substances have done what they can for the plants, new matter should be introduced to keep the life cycles going. The point with direct and simultaneous addition of organic matter to soil at planting time is to keep it moderate. Of course when you plant your crop in soil that has spent some weeks in the company of decaying matter it has already been enriched by the partial humus end-product and has the further advantage of not drawing on soil nutrients for the decomposition process. A third alternative you might like to employ is this: Make your first batch of compost a simultaneous addition to the soil at planting time, to allow you to get on with it and not waiting around for the decaying process to take place. At the same time, you can mix up a batch of organic wastes and soil to be used some time in the future. Large outdoor gardens are planned this way; as one growing and harvesting season is coming to an end, the farmer is already preparing the compost heap for the next season. This is a way of falling in to the cyclical turnings of nature and always being prepared well in advance. Another advantage will be that you can observe a comparison between the pre-decomposed method and the direct addition method. And of course it will give you practice in both.

It might be well to mention here that the suburbanite with an outdoor garden can certainly consider having an indoor one too. This is a way of having an all-year garden, much in the fashion of a greenhouse. The house-owner is also more likely to have increased indoor space to experiment with—and if he has an outdoor garden already, will be all the more familiar with gardening processes and ready to work with the necessary adjustments that indoor gardening calls for. There's no reason why an outdoor garden should preclude an indoor one—in fact, the more the merrier! The home owner, too, will have the advantage of not having to hunt out soil (he can get it right from his own backyard), and if he likes he can share the wealth of his outdoor compost heap with the plants inside. The best of both worlds! And here's another idea, incidentally—an outdoor organic gardening friend might be willing to share some of his compost with an apartment dwelling comrade! It's worth asking, at any rate.

To get more specific about the type of kitchen wastes and general organic matter you might find around your home that will be useful in your compost making experiment, we'll enlarge a bit on what we've already mentioned in our chapter on composting.

Just about any food substance that you have left over in the kitchen can theoretically be useful in a compost heap. Organic gardeners who are really purists at heart usually like to stick to the more wholesome wastes and stay away from the remains of any food that was processed in the first place. Of course with such people it is not often that one finds processed foods in their home (wise people!), so selection of wastes is not often much of a problem. Whatever other food substances you care to include, it is always good to make sure you have a good helping of the following: vegetable parings, corn cobs and stalks, nut shells and tea and coffee grounds. Outside of the kitchen, but still likely to be found in many homes: vacuum cleaner and dustpan dust, birdcage cleanings, and wood chips or shavings. If you've got some flowers in a vase that are no longer lovely to behold, chop them up and throw them in too! Remember, one of the primary rules of good composting is to insure variety. For your purposes, it is more important to make sure your compost elements are diverse than worrying about what traditional substances (such as manure) you've been unable to include. A further word on coffee grounds: these are especially good in soil that can use a boost in acid content or in soil that will be the home of acid-loving plants and vegetables. Excellent in both their high nitrogen content and their availability, coffee grounds should first be rinsed (until the water running through them is fairly clean) and left to dry to be used to their best advantage. If you've got water you just cooked vegetables in (though this is not the healthiest way to prepare vegetables!) you can use a bit of it to water your crop the next time around. Vegetable cooking water has absorbed all the water soluble vitamins from the food you cooked in it and shouldn't be discarded. In throwing it away you're throwing away a large percentage of the vitamins that food might have offered you. Proper food preparation won't avail you of this liquid, because the vegetables aren't cooked in water. But if you have cooked your food in this manner, you can still claim some vitamin benefits.

As you continue to observe and work with your garden, you'll find yourself evolving a kind of intuitive, sixth-sense awareness of its needs. When you start to have a feeling for what best suits your various plants in terms of organic wastes, you're really tuning in on nature's wavelength! This feeling for the movements of life and growth will be one of the most exciting results of your apartment gardening venture. So don't be afraid to play it by ear now and again, even in the beginning. If you keep the basics of organic logic in mind, you're not likely to go too far afield in your efforts. And if you do, simply chalk it up to trial and error and be prepared to try something different next time.

We've mentioned two variations on the compost theme that are conducive to indoor gardening. But perhaps these from-scratch processes present too much of a problem in terms of time and/or effort for the less ambitious (but still eager) at-home gardener. If this is the case, you can locate a store that will be able to sell you some ready-made humus. This is perfectly okay if you feel certain that it's a purely natural product you can trust. You might have to travel a bit to locate a store that deals in organic gardening supplies, but as the movement back to natural food growing becomes larger and more widespread, these stores are becoming more popular. At a good gardening supply outlet you'll be able to locate not only natural fertilizers, but also organic soil conditioners and mulching materials of all sorts at reasonable prices. Some of these materials available at most stores that carry organic goods are: bone meal, blood meal, cottonseed meal, raw limestone, soybean meal, cocoa bean shells, shredded pine bark, and many other varied types. All these can be used to excellent advantage in any gardening venture along with regular ready-made humus. You needn't think of store bought gardening supplies as necessarily chemical or artificial, but when you are in the market do be careful of any high-pressure appeal to buy chemical fertilizer or other products. You have just as much at stake in your small indoor garden as you would if you had fifty acres— this is your only gardening project and you should make it as healthy and pure an endeavor as you can! Chemical supplies might seem more readily available (and indeed at the time of this writing they are) and more highly advertised, but be wary and make your own decisions, based on your knowledge of the risks involved with these seemingly more convenient materials. Keep in mind that you don't have to go through a long, involved compost project in order to give your garden a rich humus application. But you might want to experiment with these methods of composting just for fun or your own knowledge and edification. The thing of primary importance is that you provide your garden plot with healthful, natural nourishment in a manner most practical for you.

The idea of the organic matter added to growing soil, whether it be the type you prepare from organic wastes around the house over a period of weeks, the type you add directly to the soil, or the kind you purchase at a store, is to benefit the soil in the myriad ways organic matter can. Even though you're working in a confined space, all the advantages of large-scale organic material application are available to you. Although your plot of "land" is a small one, because it is a capsulized version of all of nature's processes, it needs to be just as healthfully attended to as a large outdoor plot. Granted your indoor attempts at maintaining strong, vital soil won't demand nearly as much of your time as an outdoor garden, yet certain soil attributes should be worked toward. Because, just as all of the workings of the fertility chain are active in even the tiniest flowerpot, so an interruption of these by unhealthy conditions or chemical additives can be harmful to the overall project. Just as we have discussed concerning the master plan of nature, it is necessary to maintain a spongy, fibrous soil that will allow air and water circulation for optimum fertility chain functioning. Tightly-packed soil, even in the smallest of garden plots, is to be avoided, and the best guarantee you can have against it is the addition of organic matter, in just about any form.

An excellent boon to the indoor gardener (and also to the outdoor gardener who chooses to start his seeds germinating indoors in a flat) is vermiculite. Available in most good gardening supply centers, vermiculite is actually heat expanded mica, which is a form of rock. It is perfectly in tune with the organic method because it is a naturally occurring mineral without the application of chemicals. As such it has an excellent capacity not only for starting seeds but for improving the air and water holding capacities of soil. Vermiculite will actually hold several times its own weight of water and will continue to provide excellent air circulation even when it is soaked through. Because it permits this circulation of air to the plant roots it will prevent damping-off. Damping off is a phenomenon which involves the early death of young seedlings as soon as they emerge from the soil's surface. It occurs when fungi (of the destructive variety) attack the young plants at the soil line, causing them to break off at that point. It is often the result of improper ventilation—and this is where vermiculite comes to the aid of the young plant—by providing this all-important circulation of water and air. Plants troubled by damping-off are most likely in contact with contaminated soil. This can be remedied by sterilizing the soil with a bath of boiling water. Soil which has been thus sterilized should be allowed to sit for at least 24 hours, and then a ½ inch layer of vermiculite should be added loosely. Once this is done the soil should be on its way back to health.

You can germinate seeds in vermiculite and nothing else—in fact such seeds have a much better chance of avoiding stunting by insufficient water, improper temperatures or lack of good air circulation. And the whole point of germinating seeds is to provide just such a healthful, controlled environment. However, seeds germinated strictly in vermiculite must be transplanted into soil as soon as they are large enough to handle. To circumvent this need for transplanting you can use vermiculite as an additive to soil, rather than as the sole ingredient for early seed growth.

Just mix in a ½ inch or so of vermiculite on top of a mixture of soil and compost—this is ideal for good seed germination. Even before they sprout, seeds will send a long root through the vermiculite into the soil below. It is at this point that the vermiculite will help them with increased ventilation and water circulation—strengthening them for the toll-taking process of transplanting. Vermiculite is light in weight, and this is another asset in its germinating assistance qualities. You can also use vermiculite to store bulbs and winter vegetables, or even toss some into the bottom of your flower pot.

Keep in mind that with all the excellent properties of vermiculite, it does not supply any plant-growth substances. That is, it is a non-nutritive element which should never be substituted for humus or any other food needs of your growing plants. It is not as good for mulching as organic matter and you should avoid using it for this purpose. If you find yourself in a situation where you've got seeds growing strictly in vermiculite and they cannot be transplanted immediately, they should be fed a liquid fertilizer that is high in nitrogen. Continue this until you are able to get them into soil with some organic matter nutrients available.

Now that you've got your raw materials—soil, compost or organic matter in some form of decay, and the valuable vermiculite you can start to think about the proportions you'll want to use. The ideal growing ratio of these elements can be approximately one part organic matter to three parts soil, with a good solid half-inch vermiculite layer to top it off. It is possible to purchase a commercial mixture of the three, but we suggest you do the combining yourself for several reasons. Primarily, of course, it's best not to have to trust commercial pre-packaged goods—just to make sure exactly what kind of substances you'll be dealing with. After all, we're talking about the very basis of your garden—and you want to make sure it's the best and purest possible. Another good reason for avoiding commercial mixtures is that you'll want to vary the ratio with different plants. With some vegetables you might want some extra-rich soil available—so in mixing up the combination for these vegetables, you'll want to increase the humus to soil ratio. Other vegetables do not do well in soil that is overly rich—peppers, for instance tend to become too leafy if over-fertilized—and you wouldn't want that same increased proportion. A general rule of organic gardening is to put your own ingredients together whenever possible. However, the city indoor gardener may simply not have the time necessary—if you can be sure that the product you're getting is the most natural one possible, then go ahead and purchase this mixture, if you must. But keep in mind the possible pitfalls of convenience! We should also add that some vegetables, such as cucumbers, have a strong affinity for sandy soil, so you might want to throw some in your cucumber soil preparation.

Once you've got the soil mixture you need, you might do well to look around for some gravel. This can be used in the bottom of the container you'll be gardening in to insure good water drainage. Tightly packed earth will cause the water to remain in the soil longer than it should—this water will eventually stagnate and cause the plant's roots to rot as a result of soil soured by stagnant water. Fish tank gravel, for instance, is an excellent bottom layer to promote proper drainage. It has the added advantage of being inexpensive and easily obtainable at your local five-and-ten.

The type of structure you're going to need to contain your indoor garden will depend on what and how much you're going to be growing. If you've got a single fruit tree in mind, then you can work with the traditional red clay pot, which has the excellent characteristic of permitting good drainage and adequate air circulation. You can start such single plants in old coffee cans—at the time that transplanting is necessary you can simply remove the bottom of the container with a can opener and carefully loosen the dirt—pushing the entire structure out through the bottom of the can. You can also start seedlings in a milk carton cut down to a third of its height. At transplanting time you simply cut off the cardboard sides. In terms of indoor transplanting, it's a good rule to remember that you should never allow your plant's height to exceed four times the diameter of its pot. And when you do transplant, make certain to maintain the same soil level as the original pot contained. When replanting outside, it is sometimes necessary to alter this relationship of plant to soil, but indoors it is best to keep it the same.

You can actually see a tomato plant through its entire growth in a 2 gallon container, provided that you keep it in a fairly well-lit spot. It is not necessary to involve yourself with transplanting when you're working indoors. You can plant your seeds in containers large enough to hold them through their entire life span. This, of course, is most convenient when you are going to actually set up a "plot" inside your home. This is indeed a different story than having a single plant in a single pot. Such a trough can be constructed quite simply and inexpensively. You need only go to your local lumber supplier and tell him the dimensions of the trough you'd like to build. You won't need anything fancy or extravagant, and anyone around the house who can wield a hammer and nail will be able to set up a trough for you in no time. Just keep in mind that what you are trying to create is a container for a kind of facsimile "plot of ground". This is what you will need if you are interested in planting rows of vegetables or more than just a single plant in a single pot.

There are several factors which will determine the design of your plant growing trough (or troughs). First you must be alert to the needs of the various plants for space. In our discussion of specific vegetables we have made it a point to let you know just which plants are space-savers and which others are unlikely candidates for indoor planting due to their need for a good bit of room. You'll need to think in terms of the need for space between plants in a row and also the need for space between the rows themselves, in order to tabulate how many plants you'll be able to grow in your allotted space. But, in any case, plan to construct the largest trough you feel you have room for—and once you've got the exact dimensions figured out, you can go on to decide how to use the space most profitably.

The exact measurements as to width and length can be figured out from the dimensions of the room you are working in. If you've got a particularly long wall, perhaps you'd like to consider having a long trough running the distance of that empty space. It might only hold one row of vegetables, but what a pleasant and healthful way to utilize that extra area. If you can clear a big, empty space somewhere in your room you can think in terms of a square or rectangular trough of larger space area. There is also the possibility of using several troughs of different sizes to make use of a nook here or a cranny there. However, any large scale diffusion of your indoor garden is liable to result in a lighting problem— but we'll speak more specifically on that problem later.

The question of trough width and length is pretty much an arbitrary one—you can come up with the answer yourself. However, depth measurement is something else, and here you'll have to be sensitive to the needs of the plant's roots. You should, in any case, figure on at least 2 feet in depth—and more than that if you can swing it. Plants can send their roots down surprisingly far into the soil and the more room you have to accommodate them, the better.

You can think of your indoor gardening trough as a kind of huge indoor flat—that is an environment with the very best of controlled conditions for the germination of seeds and the growth of the young plants that will follow. You are master of the elements in this particular environment—you don't have to worry about torrential rains or an unexpected frost. In terms of temperature, however, you should remember that a rather moist and dry climate is most beneficial to plants. Try to keep your garden away from anything that will dry it out excessively and be constantly on the lookout for parched soil. Strangely enough, it is usually overwatering that is responsible for unhealthy plants—too many gardeners are just over-conscientious and tend to drown their plants. Later in this book you'll find a convenient list of symptoms and treatment for plant ills. This will be a helpful guide in letting you know where you might be going wrong.

LIGHTING

The presence of light is, of course, very necessary for plant and vegetable growth. There are a multitude of reasons why this is true. Light is a necessary factor in the creation of plant food for the sustenance of the plant itself. It joins with chlorophyll to manufacture food in the plant's leaves and green parts. It is this food which nourishes the plant until you have harvested it and used it to nourish yourself! Without light plant growth would not continue because there would be nothing for the plant to live on. At least some measure of light is necessary to keep the plant's vital processes functioning.

Light is often a source of warmth for the soil, as well. Indeed, the rays of the sun are responsible for the warming of the earth to the point where growth can take place This warmth keeps the fertility chain processes functioning at optimum level—it spurs bacteria growth, paving the way for the decomposition and decay activities. Temperature plays a vital role in the growth process and the sun is regarded as the very giver of life.

So, when you're setting up your indoor gardening, one of your first considerations will be how to provide proper light. Many people are not aware that a plant can spend its entire life, quite happy and prosperous, in solely artificial light. In fact, any lamp-lit corner can be a potential gardening spot and that light needn't be fluorescent, either. It is possible to grow a crop under the light of a 75 or 100 watt incandescent light bulb. If you are using this type of lighting, it's a good idea to keep the bulb at least one or two feet away from the plant, making sure that the entire general area of the plant is bathed in light. The incandescent set-up is practical with single plants or very small troughs. It's not usually possible to sufficiently light a large trough with incandescent light bulbs because the light is simply not diffused enough.

For these larger troughs you'll want to get yourself a fluorescent tube of some kind. But if you are planning to use a table lamp in your gardening scheme, make certain that the bulb is not so close to the lamp that it will dry out the leaves. Also keep in mind that the heat from light will cause quicker evaporation of moisture and be on the look-out for a possible drying out of the plant's soil. Under artificial light, soil takes on a slightly different color which may make it appear moist when it is dry and in need of watering—this is one more factor to keep in mind when you're utilizing incandescent or fluorescent lighting.

If the incandescent lighting method seems impractical or inconvenient to you, you can turn to fluorescent lighting for indoor gardening. It has the advantage of being considerably cheaper to run, and can also light a much more diffused area adequately. You can even garden in your closet with the help of fluorescent lighting!

There's a special fluorescent bulb on the market called Gro-Lux which is designed specifically for indoor gardening. Gro-Lux combines just the right properties of all fluorescent systems for the growing process. If you can't locate a Gro-Lux, or you'd rather accomplish your lighting less expensively, you can purchase a combination of red and blue fluorescent lights. This red-blue mixture will provide an excellent lighting system for your plants. You can line several plants under a fluorescent tube set-up, or extend one over your gardening trough to light different rows of vegetables. But don't forget that fluorescent tubes have a tendency to be stronger in light in the middle of the tube than at either end. With this in mind, you might want to align your trough and light tube so that light-loving plants don't get stuck in the "shade". At either end of the tube, you could place plants that are less needful of light in their growing processes.

Wherever there's a good amount of light in your home, you can have something growing. In fact, strong, direct sunlight is not always a plant's preference. This is especially true of young seedlings. Prolonged, intense sunlight is rarely good for a plant during its initial weeks of growth. When they are growing outdoors in a totally natural situation, such plants spend their early life in the shade of more mature plants. These young plants are just not strong enough to bear the often harsh rays of the sun. There are other plants which won't grow to their utmost in glaring light—and often when leaves seem parched and dry it is from too much sun exposure.

As long as you've got a reasonable amount of sunlight, incandescent lamps or some fluorescent lighting in your home, the problem of lighting your indoor garden shouldn't be a serious one. As you become increasingly familiar with your plants and their specific needs, you'll be able to make the necessary lighting adjustments. Once again, this is often a matter of careful trial and error.

When you've got your trough built or your plant pots chosen, when you've made proper arrangements for sufficient lighting, and when your soil mixture—usually 1 part humus to 3 parts soil with a vermiculite addition—is ready to go, your actual gardening can begin.

Place the soil mixture in your container, trying to get a good two feet in a trough, if that is what you are using. But no matter what holder you've chosen, you'll want to make sure that the soil is loose and as fine in texture as possible.

Trough gardeners who will be planting in rows can follow the instructions we've laid out for outdoor gardeners—the process is the same with the proper adjustments for space. You might like to consider using the broadcasting method of seed sowing we've outlined—but this is likely to call for more thinning work as the plants grow. In many cases you'll be putting young plants, rather than seeds into the soil—and we've included the appropriate instructions for this in our What to Plant chapter. If you're doing all your gardening indoors, you can make the choice of using a separate germinating pan or not. With single fruit plants, this is often a good idea; but because you'll be controlling your garden's environment all the way through, you can plant right into the soil and only transplant when the need for more room becomes apparent. Often this removal for size reasons won't be necessary and you'll be able to stick to one container for the entire life of your plant.

It's a good idea to water your seeds or young plants once they've been placed in the soil at the proper depth. Any other specific details you'll need are contained in the What to Plant chapter. For the most part, indoor gardening is a matter of ingenuity and creative adjustment. Once you know the basics of organic gardening, you have only to apply these to your own situation. With the hints we've given you as to adaptation of major organic processes, this should not be difficult. You're bound to find indoor gardening an exciting and innovative experience!!

ROOFTOP GARDENING

If you live in an apartment building which has a fairly spacious roof that is available to you, you might want to consider a co-operative gardening project with the other tenants in your building. Rooftop gardening will mean that you're dependent on the good graces of the weather in a way that you're not when gardening indoors, but there are some distinct advantages, too.

You'll need to construct a trough for outdoor gardening on your roof—and you'll have to transport all the materials too. But you'll have the benefits of natural sunlight and rains—which will cut down somewhat on the maintenance that you'll have to provide. At the same, time, however, you'll have to be on guard for frosts and storms that are likely to act harshly on your garden.

For rooftop planting, observe the suggested planting dates in the What to Plant section of this book. You can make good use of successive planting arrangements, being careful to observe the seasonal necessities of this set-up.

You could conceivably use rooftop space for an outdoor compost bin, following the instructions we've given in the composting information. Steel drums are perfect for this, and if you're able to locate a spot that won't offend anyone (including your superintendent!), then you can get started in the fall and be ready for spring planting. In fact, a rooftop compost bin can be an asset to the indoor gardener who's got a crop growing in his living room.

So, even if you're not gardening on your roof, consider it as a spot for a compost bin. You might even encourage some non-gardening neighbors to contribute to the wastes in it, if you feel the need of extra material. Remember, though, that it's a good thing to check out the acceptability of this—there's a good chance that many tenants and building owners would not be pleased with the idea! You might point out to dissenters that the compost heap is not particularly odorous once it gets going and it isn't likely to be much of a nuisance to anyone.

Rooftop gardening has the decided advantage of more space, which means less restrictions in terms of what you can grow. By constructing a large trough on your roof, you can grow as many vegetables as a small outdoor garden would provide you with. And you'll be able to grow some of the more space-loving vegetables that indoor gardening won't permit.

Cooperative rooftop gardening—working along with neighbors to grow an organic garden—is likely to be an interesting and enjoyable experience. If you know people nearby who are concerned about eating healthy food and think they could enjoy spending some time with gardening, perhaps you could work out a plan with them. Cooperative gardening has the benefits of shared work and expense—and on your roof, it's likely that you'll be able to have a yield big enough to satisfy a group of people. Cooperative work is also an excellent way of spreading the good word about organic gardening—and once you've got a booming garden on your own rooftop, you're liable to start a neighborhood fad! That's goodwill ecology that should make you proud.

If you don't have available roof space, but you've got a tiny terrace or some kind of platform outside your window—you can set up outdoor gardening there. Of course there'll be space limitations, but you can garden indoors and out during the growing season, just for variety!

If you let your imagination be your guide in seeking out gardening spots, you just might come up with something that no one has thought of before. If you keep the faith and remember that organic gardening is an experience everyone can participate in to some extent, then you're bound to find a way. Two of your most valuable gardening tools are persistence and enthusiasm!

We've devoted the two next chapters to vegetables and fruits you can grow indoors, or on your roof. They will provide a variation in method and result for the ambitious organic gardener.

WHAT TO PLANT

What shall I plant? Do I have enough room for a potato crop? Do all vegetables get planted in the same fashion? How deep shall I plant my celery seeds? What about transplanting seedlings that I've started indoors? What plants are best for a small, indoor garden?

No doubt, some or all of these questions are concerning you as you think about tackling the actual planting of your garden. In the following chapter, we've gone into a good deal of detail on each popular vegetable which you might like to think about planting. Read through it and decide which kinds of crops will make the best garden you can have. We've tried to indicate which vegetables require a good deal of nourishment, space and work on your part.

When you consider planting your garden, there are, of course, various factors which will influence your choice of vegetables. On the following pages you'll find information about many different types of vegetables, listing all kinds of necessary facts that will have bearing on your gardening plans and guide you well in constructing a basic scheme.

Asparagus

It's best to plant asparagus early in the season in soil that is rich and deep. Before planting asparagus, a heavy application of lime is necessary covered with about an inch deep layer of compost humus. Well-rotted manure should be used and the use of raw manure strictly avoided.

Plant asparagus seeds about ½ inch down into the soil, figuring on at least a 10-inch root spread. Asparagus requires 18 inches between plants and at least five feet between rows. It will take one or two years before the roots are ready to be transplanted to a permanent location. At that time they can be planted in a trench about 14 inches wide and 14 inches deep. The bottom of this trench should contain a one-inch layer of brushed limestone—be sure to mix the limestone well into the subsoil. Compost should then be added. At this point the roots can be set about 18 inches apart and the trench filled in with compost humus that is firmly patted.

Plant your asparagus in a corner of the garden, or along one side where their continued existence will not interfere with the rest of the garden's cultivation processes. During the first winter that the plants are in the ground, the area should be mulched with straw for protection. During the following spring, remove the mulch and rake the ground well. From that spring (the one following initial planting) until midsummer of the same year, asparagus shoots may be harvested. Just remember the importance of limestone in the growing of asparagus; it's highly necessary to facilitate nutrient availability.

Beets

Beets require soil that is loose, clean and even slightly sandy. The site they're planted in should be capable of good drainage and relatively sunny at all times. Early planting of beets can be at a depth of one inch and later planting at a two-inch depth. Keep a one-foot distance between rows, with about ½ foot between the plants in each row. Prior to planting, make sure plenty of manure compost is dug into the ground. Beet taproots penetrate several feet into the ground, so emphasizing a surface placement of nutrients will do these roots little good.

The ground for beet planting should be deeply dug and well prepared to accommodate the deep reaching roots. If the ground is not prepared sufficiently the roots will be starved for nutrients and the plants will not be able to survive, or if they do, will be weak and unhealthy.

Before planting beets, it is best to soak the seeds for 24 hours in water that has been poured through a container holding sifted compost humus and allowed to stand overnight.

Once you've gotten your best seed into the ground, be sure it is adequately watered. This will insure germination, and, if the weather chances to be abnormally dry at the time of planting, you would be wise to place burlap or boards over the planted seed to keep the moisture in after watering. When the seeds begin to sprout, be sure to remove the boards or burlap covering immediately.

Beets can be planted early in the spring, around March or April or again during the April to June period. If you're interested in succession planting, which involves planting a hardy crop early, harvesting it and at the same time setting a new crop of a different variety in its place, you can follow beets with late broccoli or cabbage. When you harvest early beets (which require about a 60 day maturing period) plan to use them just before maturity. Beets that are of the winter storage variety (requiring an 80 day maturing period) can be left in the ground until the first fall frosts arrive. To store beets, remove the tops, leaving about an inch of stalk, being careful not to break the skin. You can then remove the roots and store them in a root cellar.

When thinning beets, you may use the thinnings to form new rows in the garden. If the thinnings seem mature enough, they are fine for kitchen use, cooking the leaves and small roots just as they are.

Beets are excellent indicators of soil acidity. This is because they are unable to tolerate a highly acid soil. If your beets won't prosper and you've given them everything that seems necessary to a healthy crop, then there's a good chance that your soil is simply too high in acid content. This can be regulated and corrected with a liberal application of limestone. If you can't locate the ideal crushed limestone, a two-inch layer of hydrated lime can be deeply dug into the soil just before planting time. (Of course to allow for this procedure, you must be aware that your soil is acid; you can always do this as a provision for that possibility, even if you're not sure it will be necessary.)

Broccoli

Broccoli is best planted in a cool location which is open, with access to a good drainage system. It requires rich, moist, loamy soil and has a considerable need for natural calcium. You can give your broccoli the nutrients it will need by taking care to apply a ½-inch deep layer of crushed limestone the fall previous to spring planting or applying hydrated lime early in the spring. Both types of lime should be raked well into the soil's surface.

Plant broccoli seeds at a depth of ½ inch. If you're removing broccoli plants out of a germination flat, put the plants in the earth at a depth of three inches. You'll need 18 inches between early plants and almost three feet between later ones. Leave two feet between early rows and a space of three feet between late variety rows.

If you're going to be transplanting broccoli, be sure to allow for a temporary shading set-up by using inverted boxes, baskets or the like, especially if the weather is particularly hot.

Young broccoli plants should be put out in holes that are large enough to provide room for a supply of sifted compost humus and still leave room for the roots to spread out. Once the plants are in the soil, they should be well watered and firmed. The sifted compost can be placed around the roots of the plants—remembering that if you're going to use manure it should not be raw, but well rotted.

Broccoli requires a good deal of moisture to grow well. During dry spells, especially, make provisions for watering the plants with a hose, soaking the ground for several hours in the evening. Watering during the day is dangerous, as it may lead to a burning of the leaves, so always wait until sundown to carry out this procedure.

Plant broccoli in March or April for the early type and April and May for the later variety. Broccoli is ideal to use after early garden peas in succession planting and does well when set out in rows with low growing vegetables such as onions, and carrots.

Thinning broccoli should not be attempted until the plants have several leaves. At this point, thin broccoli so that it grows at a distance of 18 inches or so between plants.

In harvesting broccoli, remember that it has a central head as well as several side heads. If you gather some of these, more edible clusters will grow in their place! You can use these gathered sub-heads in the kitchen, and leave the central heads to sprout more small ones.

Broccoli is a delicious vegetable and considerably easier to grow than cauliflower. If you think you can provide rich enough soil and a reasonably conducive location, it's a good idea to put broccoli high on your list of vegetable priorities.

Cabbage

Like broccoli, cabbage requires fairly rich, moist soil situated in a location that is well drained. They also need a good deal of natural calcium in which to grow. Towards this end, you can apply limestone, raking it well into the surface once the ground has been dug.

Plant cabbage seeds about ½ inch into the soil, and put plants out at a depth of three inches. Early cabbage requires about a foot and a half between plants and late cabbage will need two feet. Maintain a 30-inch distance between cabbage rows for best results.

Set cabbage plants in holes wide enough to leave room to spread out the roots. Application of water will ensure proper contact between rootlets and soil if you are transplanting from a germination flat. Without this very necessary contact, the plants will not take root properly and will not be sufficiently nourished with soil elements. Protect late cabbage plants from the too-strong rays of the sun with inverted boxes, baskets or paper bags. Compost humus can be placed around the roots of plants, being sure to employ only well-rotted manure.

To insure the best growth possible for your cabbage, see that they are well watered. Especially during dry spells, they require a good deal of water for rapid growth and good taste.

Plant early cabbage varieties from April through May 15, and the later types from June 15 to July 30. Late cabbage varieties are good for succession planting following early peas or sweet corn and do well in companion planting set-ups (alternate rows) with radishes, lettuce and similar vegetables.

Early cabbage can be harvested at the rate you want it, once you think it has grown to adequate maturity. With the late varieties, it is best to let them stand till the first fall frost.

Cabbage is often troubled by cabbage caterpillars which can be discouraged by salt water baths.

Carrots

Carrots do well in soil that is well prepared and fertile, rich and just slightly sandy. A good location for growing carrots is one that is open, well drained and sunny. Carrots are especially dependent on a good supply of well-broken-down material. If they are fairly well supplied with this, they will grow in almost any fairly reasonable type of garden soil. Make sure, however, that the soil is free from stones and well dug and pulverized. Generous additions of humus will almost guarantee that your carrot crop will do well.

Because carrot seed is so very fine, it's a good idea to mix it with some dry, well-sifted compost before planting. When you do sow carrot seeds, place them three to four inches apart, allowing one foot between rows. Carrots are an excellent starting vegetable—because they are fairly tolerant of soil conditions (especially if you provide them with organic matter) and because they don't take up a tremendous amount of room. The seeds should be planted about ¼ inch into the soil—remember that the smaller, lighter seeds can take their place higher up in the soil structure and still do very well.

Carrots require good moisture and during dry spells it's a good idea to supply them with the water they need via a garden hose— actually flooding the area. Seed carrots thickly—you can thin them when they have attained approximately the size of a man's thumb. At that point the roots can be used as a kitchen delicacy!

Plant early carrot varieties in the March-April period and the later type between April and May. Companion planting is a specialty of the carrot—for excellent results, plant them in rows with cabbage, Brussels sprouts and even staked tomato plants. The way these plants work together will be to the advantage of them all. You can also plant carrots between rows of radishes and lettuce.

You can store carrots by digging a hole about a foot deep in a well-drained location. Place a wooden box in the hole, filling the bottom with a few inches of straw. Then put the carrots in the box on top of the straw and lay another six-inch layer of straw on top of the vegetables. Invert another box on top of this one and cover the entire structure with about a foot of dry weeds. Then cover the weeded structure with ½ foot of topsoil. In this way you've created a tiny storage cellar of your own.

Cauliflower

Rich, moist, loamy soil located in an open and unshaded spot is the best home for cauliflower. To grow cauliflower most successfully, you'll need a good deal of organic fertilizer and you can place a layer of compost along the lines where the plants will be to provide this. Don't work the compost in too well—a light job of digging should suffice.

Crushed limestone will rectify any calcium deficiency your soil might have. It is well to correct this before planting cauliflower. Again, if you can't get crushed limestone, substitute the hydrated variety and rake it into the soil's surface as early as possible.

Plant cauliflower seeds ¼ inch into the ground and the plants slightly deeper than they were in the seed bed in which you germinated them. They should be at least 20 inches apart in their rows and these rows should be separated by a 30-inch measurement.

Cauliflower is a cool weather crop—so you can sow your seeds indoors early in the spring or during the summer out in the open. When they've reached the age of five weeks, transplant cauliflower seedlings into their permanent location. Be careful not to disturb the roots in the process of transplanting the seedlings, leaving a good-sized ball of earth around the roots when you remove them. Cauliflower is a rapidly growing vegetable and due to this characteristic needs an abundance of moisture and plant nutrients at all times. After transplanting, the seedlings should be well watered and also shaded with the inverted box method. Through the entire growing period, cauliflower should never be permitted to become dried out.

You can start your cauliflower crop indoors in the early part of March, if you like, and this will give you a spring crop, once you transplant them outdoors five weeks hence. For a fall crop, plant seed outdoors in June or July. Succession planting for cauliflower can include a follow-up to early lettuce, radishes and similar early vegetables. For companion cropping, plant onions in the same rows with cauliflower or any other low-growing vegetable for good results.

There's a procedure you must follow about a week before the cauliflower plant is ready for harvesting. When the formation of a small flower head is evident, the leaves of the plant should be bent and tied to allow for a darkening of the head. This will take about a week, at the end of which the head is ready for cutting.

Celery

Celery requires very fertile soil and a good deal of moisture. It is best planted on a level site, which is also open and well drained. Celery requires a lot of plant nutrients and moisture which will render these substances available. Very acid soil will not healthfully harbor celery, and if you suspect this condition in your soil, be sure to apply large quantities of crushed limestone.

You can plant celery seeds by sprinkling them onto the earth and covering them with a light coating of washed sand. Transplanted seedlings should be set in the earth at the same depth they were in their germination beds, and no deeper.

Put celery in the ground with six inches between plants per row and a distance of two feet between rows. Start out by planting the seed thinly as soon as the ground is able to be worked, and plan to transplant seedlings in early July. For transplanting purposes, the soil in the new location should be heavily composted and well prepared. Upon transplanting, provide the seedlings with a good amount of water and provide a steady supply of moisture for the best growth results.

You can begin to harvest celery when the plants are at about ⅔ maturity.

Chinese Cabbage

Chinese cabbage thrives best on a northern slope of slight elevation; or in an area that has a tendency toward coolness. The soil should be rich and moist and the location a fairly open one. Along with a good deal of moisture, the factor of coolness is necessary to make the plant food that Chinese cabbage require available to it for assimilation purposes. You can dig a shallow trench along the line of plants and place in it about three inches of compost humus or else apply a six-inch layer of manure to the ground before planting. Either of these procedures should supply the vegetable with the plant food it will require.

Use limestone with Chinese cabbage, raking it well into the soil, or if you have time, digging a ½-inch layer in the year before planting for best results, especially in acid-oriented soils.

Chinese cabbage seeds can go into the ground at a depth of ½ inch, keeping a 16-inch distance between plants and a 30-inch distance between rows.

When planting the seeds, you can do so by digging a one-inch trench and filling it half full with composted humus. Then set the seeds in on top and cover with more sifted compost. At this point, you should water the area well.

During hot weather, it's a good idea to drench your Chinese cabbage plant area well every evening if possible.

You can provide yourself with two crops of Chinese cabbage by planting a spring crop which will mature by the end of July and following it with a fall crop to be harvested in the autumn.

Thinnings from Chinese cabbage rows may be used to form new rows in your garden, especially if moisture is provided and the plants are well shaded. Sunlight is not particularly a friend of Chinese cabbage and you should make certain that these plants are not exposed to much intense light. When thinning the plants, if you wish to use the thinnings for further planting, try to keep a good amount of soil around the roots to provide for a modicum of root hair and root system damage.

When cabbage heads reach maturity they can be cut. The most valuable and tasty part of the cabbage is the center which is delicious in salads or any other use in the kitchen. If you wish to preserve the cabbage heads, place them untrimmed in your refrigerator where they will stay for about a month.

Cucumbers

Cucumbers prefer soil that is warm, rich and rather sandy. They do best in a location that is cool and moist and even shaded. They are at home in soil that is moderately acid, but will not do well in a soil of very high acid content.

Cucumbers demand a high degree of moisture and readily assimilable plant nutrients. You can provide them with these vital necessities in a variety of ways.

One method is to dig a hole about two feet deep and two feet in diameter. Place about a pound of mature compost in the hole and water it. Then cover this area with topsoil to the extent that a slight mound is evident, rising about the ground level. In the mound, plant 10 or so cucumber seeds and thin down to five plants once they reach a height of four inches. This is called planting in hills.

You can plant cucumbers in rows if you start out by digging a trench about eight inches deep and placing about two inches of sifted compost at the bottom. On top of this layer, add topsoil and plant the seeds in a ridge in this topsoil layer.

It's a good idea to mulch cucumbers. For this job you can use dry straw, weeds, or even boards or burlap, laying these materials in about three-inch layers around the plants.

An added need of cucumbers is that of support. Cucumbers will climb a trellis and produce a crop clear of the ground if you provide them with the means to do so. Construct an inverted V trellis about five feet high and with a base of about four feet. As they grow, the vines of the cucumbers will climb these supports and produce a good, hearty crop.

May is the best month for planting cucumbers—most varieties have a maturing period of about 55 days. They can be harvested at almost any stage of their development, as long as it is before they turn a yellow-green color. The best color for table use is a deep green shade.

It's important to remember that an abundant supply of moisture is one of the major prerequisites of good cucumber growing. Don't let your cucumber plants go dry!

The most common method of cucumber storage is by pickling.

Dill

Dill is not a choosy plant to grow. In fact it will do quite well in any average garden soil, though it is best to have it located in a site which is sunny and free from tree roots. Dill requires only the smallest amount of organic material in the form of compost to nourish itself very adequately, and you can make this available by just the slightest addition of organic matter to the soil. Dill will fare pretty well in soil of a slightly acid nature, although very sour soil should be corrected by the application of crushed limestone.

Plant dill seed at a ½-inch depth leaving about a foot between plants in the row. Keep a distance of about three feet between rows. When dill plants are about five inches tall, they should be thinned to a distance of one foot apart. You can use the thinnings to form new rows or to fill up any gaps in the original row.

If you're transplanting your dill plants keep them moist and shaded during the process and once they're replanted keep them well watered and firmed.

You can harvest dill when the seed heads have developed. At this point they can be used for flavoring. You can dry them in the shade for this purpose. Use young dill stems in a salad for a delicious treat that might be new to you! Dill is a very hardy little plant. Seeds planted in the fall are often quite capable of surviving the winter and growing to maturity during the following spring. You can store dill by gathering the entire plant and hanging it in bunches until it's dry. Dried stalks such as these are excellent for use in making dill pickles out of those cucumbers you've been wondering what to do with.

Eggplant

Eggplant needs very high quality, rich soil to grow in. It's spot should be well drained, sunny and basically warm. Eggplants are very demanding in terms of the nutrients they need. Mature compost humus will provide them with the best of everything they require, and failing this, you can use well-rotted manure. Good eggplant soil tends to the slightly acid side, although not too much so, of course.

Plant eggplant at a ⅓-inch depth in the soil, and transplant seedlings from flats into a permanent location at a slightly deeper set. Leave four feet between plants in a row and another four feet between each row.

For best results, plant eggplant in a six-inch deep trench which is filled with good quality compost humus. You can plant the seeds by first mixing them with dry, finely sifted compost material. Then lay down the seed-compost mixture and cover it with still another layer of sifted compost. For setting already germinated seedlings into the ground, dig a hole large enough to permit about two pounds of sifted compost to be placed in it once the plants have been set down. Also make sure there is a good deal of room for the roots to be spread out in and that the area is sufficiently watered. At least 24 hours of shade should be provided immediately after replanting, to guard against wilting of the seedling.

Eggplants are best planted in May and do well mixed in with tomato rows. You can accomplish this by substituting every fourth or fifth tomato with an eggplant. You can also plant them alternately with peppers and expect good results for both vegetables.

Wait until the eggplant's surface is glossy before you consider harvesting it, but be sure to get to it before that surface starts to dull.

Endive

Plant your endive crop in rich, fertile soil in an open location with good drainage. The basic demand of endive is for a good nutrient supply close to the surface of the ground. You can provide this for your crop by digging in a two-inch layer of compost or doctoring faltering plants with a side dressing of compost, which at the same time will act as a mulch. Again, you can use well-rotted manure as a substitute, avoiding the raw type of manure.

Basically, endive is not too choosy about the soil it grows in, and any fairly healthy land will be able to support a successful crop. Endive should be planted in a fashion very similar to lettuce, sowing it directly into the garden and placing the seed in a shallow trench. Once placed in the trench, the seeds should be covered with a light sprinkling of sifted compost and should not be covered by more than ⅓ inch of the substance.

Keep a 12-inch distance between plants in a row and 18 inches between the rows themselves.

Endive is largely composed of water, and for this very reason it is important to see that these plants get a sufficient amount of moisture. The incorporation of humus into your soil will aid tremendously in the water retaining ability of the land and thus aid in the steady growth of the endive crop. During dry spells you might find it necessary to flood the area where your endive is growing, just to make certain that it doesn't dry out.

Customarily, endive is grown as a fall crop—it has the advantage of being very hardy to cold and can bring you a fall substitute for spring lettuce in your salad bowl. You can sow your endive seeds for a good fall crop in the June to July period. They do well in succession planting following any early crop. Harvest your endive when the heads are about 15 inches in diameter—at this point they are considered mature. You can remove the outer leaves which are liable to be soiled and withered, to get to the rich, creamy inside of the plants.

Kale

It's best to plant kale in soil which has not been the home of cabbages or related plants the previous year. Kale needs rich, fertile soil (we're beginning to realize that *all* vegetables need rich soil and compost is the quickest and most efficient way to get that soil). The best spot for kale is a site that is sunny and well drained with some shelter on the southern end. Like all other plants, kale does well with a good supply of mature compost humus or a substitute of well-rotted manure. Kale is a member of the cabbage family and, as such, requires a good application of lime for best results. You can use crushed calcium limestone or shell limestone to insure that your kale will be able to readily use the nutrients offered it by the compost you've added.

Plant kale seed about ½ inch into the soil and 15 inches apart in the row. Leave 20 inches between rows. Kale can be planted in a trench that is dug about six inches deep and six inches wide. Into this trench a layer of crushed limestone goes, and then compost. About one inch from the surface place the seed and cover this with mature compost, taking care to water well and covering the whole area with another layer of dry, sifted humus.

Plant kale around August 1—it may follow any crop which matures by midsummer. It is often considered as winter or cold weather cabbage and can be used as a substitute for it. You can sow kale in the same rows with late cabbage or storage potatoes.

Before thinning your kale crop, allow the seedlings to acquire some size so that the thinnings may be put to use in the kitchen; there's really no need to thin prior to this.

Kale is usually harvested in one fell swoop, gathering the entire crop in at one time. However, another approach is to cut outer leaves for gathering and leave the heart of the plant till it has matured sufficiently.

One of the hardiest vegetables that there is, you can plant kale late in the fall and depend on it to last all winter. You can protect it by mulching the plants with loose and heavy substances such as dry straw or weeds. After the winter, kale will resume its growth and provide you with a large and early supply of delicious greens!

Lettuce

One of the crops you're no doubt looking forward to planting and harvesting for fresh salad eating is lettuce. Lettuce is a cool weather crop and does well in rich, loose soil in a cool and temperate location. It is not fond of strong sunlight, and will grow very well in partial shade. Lettuce puts down very shallow roots and your compost work need only take an inch or so of the soil into consideration. Before planting, you can rake compost into the uppermost layer of soil and for further plant vitality add compost dressing along the rows as the plants mature. Unlike the other vegetables we've mentioned so far, lettuce can live with a raw manure application without the usual danger of disease, as long as you take care to incorporate it well with the soil during pre-planting preparations. As long as your soil is well drained, you can be fairly sure that lettuce will grow quite well on it.

Plant your lettuce seeds about ¼ inch in the ground, taking care not to cover the seed very deeply. You can put the seed in a very shallow trench to prevent its being washed away by rain. Lettuce should grow about eight inches apart in rows with a distance of a foot or so between the rows. During dry weather, take care to make sure the lettuce doesn't dry out; rapid lettuce growth depends in good part on sufficient moisture.

Early lettuce crop can be put in the ground during March and April. You'll want to plant your fall crop in the August-September period. To stagger harvest, and provide yourself with a succession of fresh greens rather than having more than you can cope with all at once, plant seeds two weeks apart until the end of May. In this way you'll not wind up throwing out lettuce that has gone bad, but will have it on hand as you require it in the kitchen.

You can begin thinning your lettuce once the young plants exhibit four leaves. If you'll be using thinnings to start new garden rows, be sure you do this work late in the day to avoid heat in the transplanting and watering processes. This, indeed, is a general rule of gardening.

Mint

Here's a treat that many gardeners miss out on—often, perhaps because they don't even think of planting it! You can grow mint right on your front lawn, as long as it is provided with a good deal of moisture and a little bit of shade. It requires a small amount of compost and you need only work this material in on the surface of your soil. A one-inch layer of composted humus will do the trick for good mint growth. You can count on mint to thrive even in moderately acid soil, too.

Mint planting is accomplished not usually by seed, but by cuttings. Mint cuttings will grow roots if placed in water, and these rooted plants are then set into the soil about ½ inch deeper than the roots themselves. Keep a distance of about three feet between transplanted mint cuttings and another three feet between rows of mint.

Mint needs a good deal of water—we find it growing naturally in very damp places and also in shady spots. Flooding your mint crop will do much to keep it healthy and prosperous; you can accomplish this by allowing your hose to run over the crop for a few hours after sundown.

There are various types of delightful mints, including peppermint and spearmint, which both have the advantage of growing quite rapidly. If you plant clippings in the spring, by fall you will have quite a good crop. Mints can be picked whenever you need them, and used for jellies, drinks and sauces as a flavoring. The stalks of mint are useful in various ways around the kitchen. Just be sure that you get the crop in before it has "gone to seed," which actually means prior to the point at which it sends up a seed stalk. At this point, mint loses its delicious aromatic value. The indoor gardener should note that mint can be grown quite successfully in window boxes all during the colder months of the year.

Mushrooms

Mushrooms will grow in any open, well-drained spot. They require an especially large amount of fairly new humus and a large supply of this is necessary for a really successful mushroom crop. If you can grow grass in a spot, then you can grow mushrooms too!

Mushrooms are propagated not by seed, but by what is known as spawn—that is, a root mass which is actually a combination of roots, mycelium and other surrounding fungoid growth. There are two basic types of spawn, one which is produced in a laboratory and is called (aptly enough) bottle spawn; the other of which is created under conditions almost natural and is known as manure spawn. To plant mushrooms in your lawn on a permanent basis, dig a hole about two feet wide and two feet deep. This hole should be filled with mature compost and then a small chunk of spawn should be placed in it. This should then be covered with about a one-inch layer of finely sifted compost.

Mushrooms will not thrive in areas which are very dry; indeed a good deal of success with mushrooms depends on maintaining moist conditions. You can harvest mushrooms as soon as they appear; if the conditions are adequate, you'll find yourself with a tremendous amount of mushrooms to gather. These can be used fresh or dried. Canning is also a possibility in preserving mushrooms.

Plant your mushroom crop during the March-April period or if planning ahead for the year to come, sow your mushrooms in September, October or November.

Onions

Onions require a damp atmosphere and soil for best results. This means soil that is moist, but not wet—proper draining, too, is a prerequisite. Good drainage implies that rain water will not gather in pools—such an occurrence is liable to mean stagnant water and rotting plant roots.

During the early stages of their development (and later on, too, though to a lesser degree), onions require a solid dose of vital plant nutrients. They're especially dependent on a strong supply of nitrogen—and a heavy layer of compost will supply this to growing onions.

Onions have a preference for soil that is moderately acid and should be planted as seeds, about ½ inch in the ground. Onion sets, or small onions can be laid into the soil at a depth of one inch, while plants should be set in slightly deeper than they grew in their flat, if transplanting is involved. You can plant onions fairly close, usually allowing about two inches between plants, although very rich soil should allow you to crowd them in even closer. Figure on leaving two to three feet between onion rows.

Plant your onion seed quite thickly, and cover with ½ inch of compost that has been sifted, if possible. Subsequent to planting see that the area is well watered. If you're planting onion sets, then increase the compost covering to one inch, being careful not to use onion sets that are larger than a dime; onions of this size and larger are likely to flower rather than to develop into an onion. If you're transplanting keep the seedlings shaded from summer heat as much as possible. You can use onions, in their capacity as a low-growing vegetable, to fill in the spaces of rows containing tall, widely spaced vegetables like late cabbage or broccoli.

As onions mature the tops of them will sag to the ground. For onions that don't droop in this way you should break them down with a rake to allow for the maturing of the bulbs. This can be further facilitated by pulling the bulbs and spreading them out on the ground in the sun for a day or two. After this, you can cut off the tops an inch above the bulb.

Once you've harvested onions, place them in a net bag in an area that will promise free air circulation. When cold weather sets in, they should be moved to a storage cellar that is cool, dark and fairly dry.

Peas

Grow your peas in light, rich, sandy soil in partial shade. Peas will put up with a moderate amount of acidity, but don't do well in soil that is overly rich in nitrogen.

Plant peas in trenches that are about one foot deep, allowing for about three inches of compost to be placed in the bottom of these trenches. The pea seed should be sown about one or two inches below a layer of sifted compost. Keep two to four inches between plants and about three feet between the rows themselves.

As the seedlings grow, the sides of the trench should be broken down to cover everything but the tips of the seedlings. Continue this process as they grow, always leaving the tips exposed, until the trench is filled in totally.

Adequate moisture and shade are vital factors in the healthy growing of a good pea crop. Heat and drought are the most common enemies of the pea crop.

For peas other than the bush variety, you must provide some support for climbing purposes. As soon as the plants begin to grow out of the ground supply some material for this purpose, fixing it securely and measuring it to be just a bit taller than the estimated mature growth of the peas.

As with lettuce you can sow peas successively about two weeks apart until the middle of May or perhaps a bit later. Once you've got your peas harvested, employ successive planting by putting some late cabbage or other late crop into their place.

You can avoid the time-consuming job of thinning if you are careful to plant no more than 12 to 15 seeds per foot. This will also cut down considerably on wasted crop. And plan to harvest your pea pods daily once they are ready—that way they won't interfere with the growth of new and developing pods.

Peppers

Peppers require very sunny, open territory with soil that is mellow but not particularly rich. Like peas, peppers do well in a soil that is not very heavy in nitrogen. Too much nutrient availability will foster a heavy leaf growth which is not favorable to the health of the peppers themselves. So humus should be moderately applied, digging perhaps an inch of it lightly into the topmost portion of soil. It is best to avoid acid soil with peppers and also to try and have a kind of gravelly texture to the earth. Pepper seeds should be planted no more than ½ inch in the soil and the plants that have been previously germinated will do well in the ground slightly deeper than they were in their flats. Keep about two feet between pepper plants in the row, and a space of 30 inches between the actual rows.

After setting down a shallow layer of compost, dig a shallow trench and drop the seed in, covering it with ½ inch of soil. When putting down plants, make sure the hole is large enough to allow the roots to spread out quite comfortably and keep the young plants protected from the direct rays of the sun. Also make sure the area is kept moist and that the plants are completely shaded for their first day or two outside.

Heavy watering of pepper plants is necessary only during the early stages of their growth—you can relax as the plants develop, because their need for water will decrease.

Plant your peppers in May or June and plan to use them as follow-up crops to garden cress, mustard, lettuce or other early plants. For companion cropping, sow onions and carrots between the pepper plants in the same rows.

Harvest your peppers when they look ripe for kitchen use. For storage you'll most likely be using them in the making of pickles and other condiments.

Potatoes

Strange as it may seem to you at first, potatoes favor a cloudy day! They are fond of mild weather, however, because their sensitivity to frost is very great. Soil of a sandy quality is best for potato growing and the use of lime in the soil should be avoided. Potatoes do best in soil that is slightly acid; too much alkalinity of the soil (a condition brought on by use of lime) will cause the rapid development of a potato disease known as scab.

If you're using potato seeds, plant them at a depth of three inches, leaving 13 inches between plants and two and a half feet between rows. For the actual planting, dig trenches about 10 inches deep and lay down a two-inch layer of sifted compost at the bottom. The seed is then placed in the trench with two inches of sifted compost on top. This is covered with enough fine topsoil to form a slight ridge.

During dry spells keep your potatoes quite moist as they enjoy moist, temperate conditions.

Early potato crop should be planted in the March-April period and can be used in successive planting with late turnips. Your late potato crop should be planted at the end of May. Between early potato rows, plant sweet corn or late cabbage for excellent results.

Gather your early potato crop in the summer. Late potatoes can remain in the ground till much later, but should be harvested before the frost sets in. You can store potatoes temporarily in a cool basement, but for more permanent storage it's a good idea to dig out a pit for the purpose. A circular area of seven or eight inches deep and six feet in diameter can be filled with dry litter and the pile of potatoes set on that. On top of the potatoes, place a foot-thick layer of dry grass and a foot-thick layer of topsoil on top of that.

Radishes

Radishes need very little plant food to thrive. They do well on light sandy soil that has excellent drainage. A one-inch layer of compost will do the trick, and you can count on radishes to grow in just about any good garden soil. Radishes should be planted at a depth of ½ inch in the soil and covered by a layer of sifted compost. A foot of garden row will take about 15 seeds quite happily, and you should leave about a foot between rows. This ranks radishes as one of the less space-consuming plants and indoor gardeners would do well to keep that in mind.

Radishes demand a tremendous amount of water to grow rapidly and healthfully; as long as you've got proper drainage availability it's virtually impossible to give a radish too much water.

You should plant radishes in successive stages every week from April 1 into early summer. Just remember that the fine flavor and crunchiness of radishes depend almost entirely on the sufficient supply of water to their roots.

Scallions

You can grow scallions successfully in the shade, but they'll respond better to a sunny area that is also open and moist. Rich, moist soil boosted with a good dose of humus will produce the best scallion crop you could hope for! Because scallions are perennials, they demand an abundance of natural plant nutrients to insure their continued reproduction year after year. Towards this end you should provide your scallion growing area with a tremendous amount of compost humus.

Plant scallions at a ½-inch depth leaving four inches between plants and 18 inches between scallion rows. Plant the seed thickly and cover it with ½ inch of sifted compost humus for best results.

If you sow your scallion seeds in the spring or summer you can harvest a crop either in the fall of that year or the spring of the following year. It is a good idea to harvest in the fall to help this perennial establish its hold in the soil. You can plant scallions after the harvesting of early carrots, beets or any other spring greens.

Even though scallions are a perennial and are really quite hardy, it's good to provide them with some protection to make it through the winter months. You can do this by covering the area with a loose mulch of dry weeds, leaves or straw. Try to keep the scallions to one side of the garden where they won't interfere with your cultivation of annual crops.

Spinach

Spinach is a plant that requires a good deal of nitrogen to prosper. This plant enjoys low temperatures and requires an abundance of plant food. A heavy layer of manure will insure that the necessary nitrogen is available to your spinach crop and you can figure on as much as a six-inch layer for this purpose. Any relatively good garden soil will make a healthy home for spinach.

Spinach seeds should be planted at ½-inch depth and covered with sifted compost. Keep them about four inches apart in their rows and 15 inches between rows. In the case of spinach it's a good idea to see that the seed is properly spaced and the soil firmed once the seed is in the ground. Firmed, by the way, does not mean tightly packed, just *firmed*.

Spinach is basically a cool weather crop and will not be damaged by late spring frosts. You can plant a row of spinach every two weeks until summer and resume that procedure in the early fall. Early spinach is planted in March-April and the late varieties in August-September. Early spinach may be followed by the planting of bush or pole beans.

Harvesting spinach is best accomplished by gathering the outer leaves and leaving the center of the plant intact to continue to grow and produce new outer leaves.

Squash

Squash will do well in sandy soils that have good drainage and warm up quickly. Mature compost humus or well-rotted manure will provide squash with the nutrients so necessary to its growth. Slightly acid soil is good for squash growth. Summer squash should be planted at a depth of ½ to one inch. Winter squash needs the whole inch. For summer squash leave about three feet between plants in a row and for the winter variety it is best to leave six to nine feet. We can see immediately, then, that squash is a space-demanding vegetable and only a gardener will want to consider it. This space is consumed mostly by the spread of the plant's vines.

Winter squashes have the advantage of being stored and brought out when other vegetables are scarce.

With squash planting avoid too-thick seeding as this will mean a good deal of work thinning later on. Harvest summer squash as you need it during the summer, but leave winter squash in the ground till just before the first fall frosts. You can store winter squash in single layers on shelves in a cool place. It's a good idea to wipe them off every few weeks to prevent the formation of damaging mold.

Sweet Corn

Corn is a very tough plant and it has the ability to thrive in places which couldn't harbor many other vegetables. It requires a fertile, medium soil that is not overly rich and it will do well on sunny hillsides which have good drainage. You may plant sweet corn in rough soil, but make sure that you supply your crop with adequate natural humus for its growth needs. You can do this by digging a one-inch layer of compost into the area where the corn will grow, substituting if you must, well-rotted manure. Corn will also require an adequate calcium supply, and if this is lacking in your soil, it may be provided by an application of limestone.

Plant your corn seeds at a depth of ½ to one inch. Early corn should be spaced about three feet apart in its rows, and make provisions to leave four feet between late variety plants. You can get away with three feet between rows for early corn, but better make it four for late types.

Don't plant your corn crop until all danger of frost has passed and the ground is starting to warm up. You can plant corn in hills if you wish, putting three seeds in each hill. And don't leave any loose seed lying about on top of or near the hill—this will attract crows that will not only eat the available, unplanted seed, but will dig into the ground to get to the seed that's planted in the hill. Early corn varieties are to be planted in the April-May period, and you can start your late crop in May-June. Stagger the planting by putting a row of sweet corn in every other week, this way you can spread your harvest over a longer period of time. Companion planting can include putting corn in rows with low growing vegetables, such as carrots or onions. If you plant head lettuce in rows with sweet corn it will profit by the shade the corn provides. This kind of mutual advantage is what companion planting is all about. When you're plotting your garden, it's always a good time to see what kind of space-saving ideas companion planting can provide you with.

You can plant popcorn, in small amounts, and get very satisfying results if you think you might enjoy having some of the home-grown type around. You can also obtain seeds for growing multi-colored corn ears, well-known for their decorative touch at fall celebrations.

Swiss Chard

Any soil that is provided with a reasonable amount of organic fertilizer will grow an admirable crop of Swiss chard. Supplying the soil with lime will be good insurance that your crop will be a strong and healthy one. You can lay the compost for your Swiss chard crop down in a small ridge along the line where the crop will stand.

Swiss chard seeds should go into the soil at a one-inch depth with eight inches between plants in a row. You need only leave about one and a half feet between the rows, which means that chard is an excellent crop for small gardens or an indoor plot. You can grow a good amount in a small amount of space.

Companion cropping with chard can include sowing it in a row along with radishes or beets. (Swiss chard, by the way, is itself a member of the beet family.) You can thin your chard crop as soon as they appear to be crowding each other and use the thinnings in the kitchen. At time of harvest, leave the inner, younger leaves to continue producing greens, and gather only the outer leaves. This way you'll have a good supply all during the summer. Chard can even produce through the winter, provided that it is not extraordinarily harsh—but cover the crop with a dry, loose mulch for protective covering.

Tomatoes

Tomatoes like light soil which is fairly porous and well drained —they're happiest in a sunny and open location and with the addition of a good amount of compost, you'll raise a hearty tomato crop.

Tomato seeds can be planted at a ½-inch depth, while the plants go best in the earth at about one inch deeper than they were in their germination flats. To discourage cutworms, you can place a three-inch square wrapping of paper in a roll around the stem of each plant slightly below surface level. Work a good amount of compost around the roots at the time they are set out. Tomatoes also respond quite well to mulching as soon as the weather becomes warm.

Leave three to four feet between plants in a tomato row and four or five feet between the rows themselves.

When you're transplanting a tomato into its permanent spot in the ground, make sure that the hole you've dug will permit the plant to spread its roots out comfortably. An application of water will provide good contact between the roots and the soil, and it is at this ime that the compost is molded around the plants. In very warm weather, young plants should be protected with inverted boxes or baskets, as we've described previously.

You won't need to water your tomato crop unless there's a particularly dry spell. The May-June period is most conducive to tomato planting, making sure that any danger of frost is over.

Larger tomatoes will require stakes. Prune your plant to one or two stems and tie them to stakes that are about the thickness of a broomstick. These stakes can be driven into the ground at a depth of about eight inches for best results.

Make certain to harvest your tomato crop prior to the frost. If this necessitates gathering some that are still green, they may be stored in a cool, dark place where they will ripen without light and provide you with tomatoes during the winter.

Turnips

For your turnip crop, you'll want sandy soil that is not particularly rich in nitrogen. Turnips are basically a cool-weather crop and they require good drainage along with adequate moisture. With this crop, you needn't worry much about a lot of extra organic fertilization. Although turnips are capable of thriving in soil that is not too rich, try to avoid an excess of acidity for this crop. An excess on either side of the alkaline/acid scale will not be beneficial to the growth of almost any plant you can think of. By now, you've no doubt made that observation on your own!

Plant turnips at a depth of ¼ to ½ inch, allowing four to six inches between plants in a row. About one and a quarter feet will suffice between rows—turnips being another plant excellent for small-scale gardening projects.

Your turnip seed can be set into a shallow trench formed by using the corner of the hoe. Sow the seed thinly and cover it with sifted compost if such is available.

Turnips send down taproots to a pretty good depth; because of their large roots they require a good deal of moisture for nourishment. You can provide them with this by giving them a thorough soaking every so often.

Any crop which matures before the end of July can be successfully followed by a turnip crop. Turnips have the capacity to nourish themselves on nutrients that other plants didn't use—so you can plant them wherever there's space and be certain that they'll grow quite well.

Turnips have an ability to withstand frost, but should not be left in the ground once it is frozen solid. You can store them in a root cellar or in a pit dug in the ground.

By now you've probably got a good idea of what vegetables will be most feasible for you to plant in terms of limitations of space, soil and available organic materials.

Vegetables such as turnips, Swiss chard, spinach, radishes and carrots are especially good for the small plot or indoor gardener in terms of being space-savers. They don't require too much space between plants and you can keep the rows fairly close together without endangering the crop. These vegetables also tend to be less particular in terms of nourishment than many other types—although you should see that your spinach crop has the requisite amount of nitrogen.

Space limitations in your garden won't stop you from planting onions, peppers, tomatoes, scallions, celery, cauliflower or many of the other vegetables we've listed that are not exactly space-savers but at the same time don't need a tremendous amount of room in which to grow.

Potatoes, peas, squash and sweet corn will present a bit more of a problem in terms of space. You can utilize our companion planting suggestions if you're eager to grow these vegetables but are somewhat short on area. Certainly don't forsake any planting ideas till you've tried to figure out if there's a way you can swing it, always keeping in mind possibilities of successive planting and companion cropping. And don't be afraid of trial and error—if you feel like experimenting and have reason to believe your idea is at least fairly feasible, then by all means go ahead. It will keep your spirit of adventure in the right place and possibly even provide you with some innovative gardening success. If your plan is a failure, then chalk it up to experience and keep plugging!

IN SUMMARY

In your perusal of our specific vegetable planting instructions, you've no doubt drawn certain conclusions which can serve as good universal rules for gardening—whether it's indoors or outdoors, be it a vast and ambitious venture or a modest, careful one.

You've learned (and certainly we've emphasized it enough!) the tremendous importance of the addition of organic fertilizers for successful home gardening. No matter what type of soil a particular plant is partial to, no matter what location it will thrive in, no matter how much space it will need to grow in, or what weather it will do best in, that plant is going to need organic matter and it's up to you to provide it! Compost, compost, and more compost —it's a guarantee for a good crop and a panacea for soil ills! Read over the chapter on compost and understand the concept one more time—refresh your mind on the myriad benefits that decayed organic matter can offer your soil and plants. Let yourself begin to think organically and to plan your gardening efforts with a good concept of the cosmos in mind; that is, begin to see yourself as a part of the very flow of nature we've been explaining. As you feel yourself more and more a part of things (and indeed you are— more than you know!), your intuition about gardening and making the proper decisions will become a strong tool at your command. After all, having a green thumb is really no more than understanding nature and giving her what she needs to function at optimum level. Remember that nature is ready and willing to do all the really difficult work, and you have only to provide a conducive environment. Indeed, the whole idea of organic farming and gardening is to stay in tune with nature, to work in cooperation with her, and at times to anticipate her needs. For instance, by realizing that calcium is a necessity, say, in the growing of a good cauliflower crop, you can provide this by an application of limestone, either crushed or hydrated. Because you know in advance that your onions are going to require a high availability of nitrogen, you see that this is made available to the crop by offering the soil a good deal of organic matter in the form of compost. As you work you'll begin to get a real feeling for the specific needs of each crop, and as you become aware of these needs, you'll be prepared to provide them. Your role as vital assistant to nature's processes does not depend on the size of your garden, or its location—not in the least! Your careful guidance and willing cooperation with the flow of life are necessities, whether you're growing a three-acre field of corn or two tomato plants in your den. We've said it before—even on the smallest scale, you can participate in the workings of life's growth processes, just by having a garden of your own—regardless of its size.

Watch your vegetables as they grow—keeping their sundry needs in mind, prepare to administer first-aid if it seems necessary. Does the soil seem too dry during a particularly hot, sunny spell? Get out the hose (after sundown!) and give the area a proper dousing. Remember, too, that unhealthy looking plants may often be bolstered by a "side-dressing" of compost. You can add organic matter to the soil, if you're quite careful not to disturb any plant or root system structures, with excellent results. It is always best to go about enriching your soil as early as possible, but in the event that some extra added help seems in order, don't hesitate to experiment in this manner.

In your consideration of work with organic matter, remember the many advantages of mulch, too. We've mentioned it in specific conjunction with certain vegetables, but it is generally an excellent method of adding organic matter (which will decay more slowly than compost which has already begun the breakdown to basics) to the soil, and keeping down weed growth at the same time. Mulch can also be employed as protection for some winter-hardy vegetables, such as turnips, once the cold begins to set in. Just like composting, mulching is very basic to the practice of organic farming and should be emphasized and worked with again and again in your garden. Like composting, it can be carried out on whatever scale seems appropriate for your gardening. But even if you have only a small crop growing, you'll want to use mulch and also enjoy the adventure of experimenting with its possibilities. The pioneer spirit of organic gardening is one of its greatest assets, and if you keep certain priorities in mind, you'll find yourself not only growing (right along with your vegetables!) but learning at the same time.

We'd like to make some relevant remarks on the general subject of harvesting. After all, what's the point of working hard at growing a strong and healthy crop if you've not been acquainted with the most beneficial harvesting rules? We've already included specifics as to methods for each vegetable that should help you in planning to reap what you have so carefully and lovingly sown.

It is a good rule of thumb that vegetables be picked at the time when they taste best to you—at the very moment that they seem to be of the highest possible quality attainable. Scientifically speaking, this will most likely be at a time just before they go through chemical changes which convert sugar to starch and fiber to cellulose. Testing your own crop for flavor and aroma can be an exciting and rewarding experience. You won't have to hope for the best as you do in the supermarket—if something tastes not quite ripe, perhaps you'll choose to leave the remaining crop in the ground for a few more days. If you prefer vegetables at an earlier stage of development, then that opportunity is open to you, too. The time of day at which you harvest has an actual and very real effect on the vitamin content of your vegetables. To receive maximum vitamin C quality, harvest your vegetables only after 10 o'clock in the morning, following fairly clear weather. It is at this point that vitamin C content will be at its height, since it actually does vary depending on the weather and amount of light available to the plant. The factor of timing in harvesting is much more important than most people realize. Below we've given a table of harvesting hints for gardeners:

Asparagus: You won't be able to harvest your asparagus crop until after the third year. At this point, the spears should be at least six inches above the ground before you bring them in.

Beans: Pick your snap beans about three weeks after they first bloom. At this point the pods will not be quite full sized. If you let them wait much longer they are liable to be tough and difficult to chew.

Broccoli: It's best to harvest broccoli before the dark green blossom clusters begin to open.

Cabbage: Wait until your cabbage heads are solid before you harvest, but try to gather them before the heads split. And don't forget those salt water baths during cabbage growth!

Carrots: Hold off on harvesting your carrots until they are an inch to an inch and a half in diameter.

Cauliflower: The outer leaves of cauliflower should be tied above the head when the curds are about two inches in diameter. The heads will be ready to harvest about 10 days later.

Sweet Corn: Wait until the kernels feel fully developed, but before they reach the doughy stage. The corn silks should be brown and rather dry and the tips of the ears filled out tightly.

Cucumbers: Catch your cucumbers while they are slim and green in color, and before the color starts to fade. If you're harvesting for pickles, leave a short piece of stem on each cucumber.

Eggplant: Gather your eggplant when they are about half grown —prior to becoming dull in color.

Onions: You can harvest your onions for table use when they are ¼ to one inch in diameter. If you're planning to store them wait until the tops fall over and shrivel at the neck of the bulb. Get them in before the heavy frost.

Peas: Harvest peas when the pods are firm and filled out, but don't let the seeds ripen to their fullest size.

Peppers: Gather when they have become solid but are still not full sized—if you're dealing with red peppers, wait until the color is uniform to harvest.

Potatoes: Potatoes can be harvested when the tubers seem large enough. If they're not quite ripe the skin will be thin and rub off easily. For potato storage, make sure the tubers are quite mature and that the vines are dead. Until the vines are dead, the tubers will continue to grow.

Squash: Summer squash varieties can be harvested in their immature stage—this is while the skin is soft and before the seeds are ripe. Winter squash, however, should be allowed to mature to the point where the skin is very hard. Get your winter squash in before the heavy winter frost comes along.

Tomatoes: Don't let tomatoes get soft before you bring them in, just make sure that their color is a uniform red.

Turnips: It's time to harvest turnips when they are about two or three inches in diameter.

Most people don't realize that waiting too late to harvest and allowing vegetables to become overripe is probably a greater and more frequent problem than gathering them prematurely. If a crop remains in the open too long it will start to deteriorate, not only in taste but in nutritional content. If you're planning to harvest for use at the dinner table the same evening, it's a good idea to wait as long as you can before bringing the crop in, to insure freshest taste and highest quality possible. If it's a salad you're planning, why not run out to the garden (or into your living room) just before dinner and gather up the greens or carrots then? It's a fun experience and will guarantee you fresh and wholesome eating. If you're planning to freeze or preserve vegetables, don't bring them in until you've got the preparations made for the preserving procedure. Picked vegetables, sitting around your kitchen, will continue to lose vitamin content, even though they are no longer out in the open.

So whatever vegetables you choose—whatever crops seem to fit in most with your family's needs and your spatial or time limitations, keep the rules of composting, mulching, watering and good harvesting in mind, and your crop will be a delicious, nutritious winner you can be proud of!

GROWING FRUIT INDOORS

Along with vegetables, you might want to consider planting some tropical fruits in your indoor garden. Your own tropical jungle can include mangos, pineapples, avocados, prickly pears, artichokes and papayas.

The soil proportions for your tropical and fruit plants can be the same as the vegetable planting ratio we mentioned. Three parts good, rich soil to one part humus or composted materials and a good-sized portion of vermiculite will do the trick. You can start these plants in milk cartons or coffee cans, carefully rinsed. Or you can construct a miniature hothouse with an aluminum baking pan—layering the bottom with four or five layers of cheese-cloth—for an excellent seed germinating set-up. If you're going to germinate your seeds in a pan of this sort, keep a sheet of plastic over the top, and try to keep the pan in a darkened spot. Seeds will remain in these "hothouses" until a taproot emerges along with a few leaves. We've already discussed the necessity of root and green leaves in permitting the plant to function independently. Once your seed has germinated in this manner, move it into a larger pot. For single plants of the tropical or fruit type, you will do well with the red clay pot we spoke of earlier. These pots come in fairly large sizes and you'll be able to accommodate your tropical plant until a point when it reaches really vast proportions.

If you'd like to plant mangos, begin by soaking a pit removed from a piece of the fruit in warmish water for about a week. Be careful to change the water each day—this will allow the seed to germinate properly and without rotting first. After the soaking, plant your mango seed (with the eye facing upward) in fairly loose, rich soil in the container you have chosen for it. At the time of planting, be careful to water the mango seed very heavily— in fact, this is a good general rule for the initial planting process. You can leave it alone for awhile after that—watering it with tepid water when it begins to look dry again. Don't make the all-too-common mistake of watering your plants too often. And when you do water the plant (and this goes for any potted plant) do it

134

slowly—making sure that you are not wetting the soil at a greater rate than it is capable of absorbing the moisture. Glutting a plant with water is not good gardening—and a plant that is over-watered will begin to show symptoms of unhealthiness. We've listed some of these symptoms in a summary at the conclusion of this chapter. Remember, too, that it is good to water from the bottom twice a month or so—you can do this by putting water in the dish you've set beneath the pot. Such a procedure will insure good water circulation top and bottom.

Mango plants grow in spurts, so don't be discouraged if you seem to see no progress for weeks at a time. After a dormant period, the plant will begin to grow once again.

Mango trees make exceedingly attractive indoor plants, and even if yours don't flower and fruit—they'll provide lots of green beauty for your indoor decor.

Pineapples are very tolerant plants—they don't dry out or wilt easily and they're pretty good at taking care of themselves. Pineapple roots do not grow deep, so you won't have to worry about a deep container or a large amount of soil mixture.

One thing pineapples are partial to, however, is a high acid content in the soil. For the indoor gardener this is no problem at all—you can simply make use of those coffee grounds you're always wondering what to do with. Just remember to wash the grounds carefully, and in the case of pineapples, dry them out before adding to your soil mixture. These coffee grounds will provide the necessary addition of acid to soil for good pineapple-growing results.

To get a pineapple plant started, you'll need the green top of a pineapple. Also, leave a bit of the fruit in with the portion you cut off—burying the entire piece in your soil, bringing soil all around the slice, right up around the crown. Again, give your plant an initial heavy watering. One of the happy parts of pineapple planting is that you don't have to germinate a seed first.

Pineapples do like sunshine, but you can grow them without a good deal of it. And you can count on your pineapple plant to fruit—but it is likely to take as long as two years or more. If you're really interested in having your plant bear fruit, then do try to give it a couple hours of direct sunlight whenever you can and try to maintain a 75 degree Fahrenheit room temperature if possible. If it is impossible for you to provide these imitation tropical conditions for your indoor growing pineapple, you may be just as pleased with simply the attractive plant, even if it doesn't bear.

Interested in *artichokes*? Try planting them in a variation of the soil proportion formula we've given—use two parts soil and one part humus. You'll want to plant the "tuber" which is similar in appearance to a potato with a lump in the middle. These are available at the grocery store—or rather, at any grocery store that will carry artichokes in the first place. Be careful to cover the tuber in at least two inches of your planting mixture, it needs protection from city air, being a highly sensitive kind of plant. By the way, an addition of sand to your soil mixture will be a boon to the artichoke and be certain that you give them lots of water—unlike many other plants, artichokes crave a good deal of moisture and will dry out quickly if deprived of it. Artichoke plants are liable to shoot up to eight feet and more—so beware!

Papayas are quick-bearing plants—that is, nine months after it has first sprouted, the papaya will bear fruit—and one plant will hold several fruits.

Plant your papayas in a terra cotta pot—this is a necessity because papayas need extraordinarily good aeration and water drainage. Standing water can kill off papaya roots in 48 hours! You can avoid the danger of root and plant rotting by mixing a fairly loose soil for your papaya plant—try half soil and half humus with a good measure of vermiculite. It's good to keep in mind that papaya plants are very sensitive to transplanting—so it's well to plant them in large containers so that you can avoid moving them until they are quite large and sturdy.

It is somewhat difficult to get a *pomegranate* plant to flower— but if you've got a really sunny spot in your apartment, you might want to try it. You can also make use of a spot that might have seemed too dry or hot for your other plants—pomegranates don't mind dryness and direct heat nearly as much as most plants.

Germinate your pomegranate seeds in a clean container layered on the bottom with moistened cheesecloth. It's likely to take about six weeks for pomegranate seeds to germinate and put out a taproot. When you've got a root, move your seed into a soil mixture that is about one part humus, two parts earth and one part sand—and don't forget the vermiculite. You need only put the seed (with the root facing down) about ¼ inch into the soil—cover it lightly and water generously. And if you are going to be keeping your pomegranate in that extra dry spot in your apartment, remember that you will need to water it—while not as sensitive to dry conditions as most plants, pomegranates still need adequate moisture to survive.

Are you partial to citrus fruits? Then you're in luck, because these tasty types are among the easiest of fruits to grow indoors. For your lemon, orange or citron seeds you'll need a soil mixed with half soil and half humus and a good handful of sand, along with the usual vermiculite. After you mix this compound, you might want to add an extra ½ inch or so of humus, just to insure that your plants get some extra nutrition which they're likely to need.

Well-distributed fluorescent lighting, as we described in our Indoor Gardening chapter, will be your best bet for supplying these light-loving plants with the brightness they require. But, remember that you can grow your citrus quite well underneath an incandescent light bulb, too. Citrus fruit plants are really quite adaptable to varying light conditions. What they do not adjust to well at all, however, is periods of extreme dryness. Water your citrus "orchard" twice weekly and also give the leaves a light misty spray each day. You can accomplish this with a spray bottle and some tepid water—taking care not to water at a time when direct sun will be on the leaves. Any bright light shining on wet leaves will tend to burn or yellow them.

Waiting for your citrus fruit to blossom and bear is going to require some patience on your part. It's not likely to happen for five years after the seed is germinated. It's well worth waiting for, we think—and in the meantime you'll have some lovely greens in your apartment!

Those are just a few of the fruits you can cultivate in your role as indoor gardener extraordinaire! Fruit growing, which can be accomplished in single pots, may be more convenient for you in terms of space than a large trough with a row of carrots. Or maybe you'll be able to cultivate both a vegetable and fruit garden—all the better and more exciting!

We've listed a couple of hints that will guide you well in your pursuit of indoor gardening on the following pages. You might want to mark these for re-reading and consultation as your gardening project progresses. After awhile, they'll become a firm part of your gardening experience, and as such, will not require any reminder.

Perhaps the first rule to keep in mind is that your intuition and common sense is likely to be your most valuable guide. As we've mentioned before, having a feeling for your garden will lead you to the best considerations possible. As your plants grow before your eyes, so will your understanding of their needs and preferences. But there are several points it's good to keep in mind:

It's best to avoid use of commercial fertilizers in your indoor gardening. Many such fertilizers used in potted plants, for instance, will burn the leaves of your plants—even if administered according to directions. In the spirit of organic gardening, we suggest you stick to organic composting and humus addition methods for the best and healthiest results.

To keep the soil in your indoor garden loose and loamy, dig around in it a bit every couple of weeks. This is especially appropriate in the case of potted plants which tend to pack soil rather tightly. You must be careful, however, not to disturb any vital root structures in this loosening procedure—even though your plants must breathe, once you've destroyed or seriously disturbed the root structure, your plant will falter and die. It is especially important to do this carefully if you are dealing with shallow rooted plants, whose vital structures are nearer to the surface and thus, more vulnerable.

Try to vary the types of organic matter you're adding to the soil—this is one of the foremost rules of successful organic gardening. Anything we've listed that's available to you, or anything you can think of yourself that is organic, will be life-giving to your indoor plants. But don't get stuck using the same type of material over and over. Just as this is one of the major guidelines for effective outdoor composting with traditional methods, it is important on a smaller scale, too.

You can use your plant pots or garden troughs as ashtrays—well, partially at least. Cigarette and cigar ashes are very high in potassium which helps in making the leaves nice and green—but tobacco is not good for your plants. So encourage guests to flick their ashes in your plants, but don't allow them to put the butt out there.

Always use tepid water for watering your plants—cold baths are not at all to the liking of your indoor greenery. And we want to emphasize, once again, the danger of over-watering. A lot of gardeners feel compelled to water their plants each and every day —this is not at all a good idea and will only lead to trouble where the plant's health is concerned. For plants that require more moisture, you might want to water more frequently—but as a rule two times a week should do it. Plants that are particularly high in their moisture requirements have been mentioned as such in our specific vegetable and fruit discussions.

Avoid "softened" water for use in your garden. There are certain chemicals used in the process of water softening that are destructive to plant growth. It is also a good idea to leave heavily chlorinated water out overnight in an open container before watering your plants with it. This will help to cut down on the effect of the chlorine on your plants.

As a rule, plants enjoy a humid atmosphere. You can simulate this by lightly spraying the leaves of your plants every so often (not too often, remember!). However, if the plant leaves curl up right after you spray them, it means that your apartment is too dry to allow for natural water evaporation. This might be somewhat destructive to your plants, so spray around the plant rather than directly on the leaves. If your plant seems to be suffering from the dryness, try moving it into the bathroom for a couple of hours a day. The environment in there is likely to be moister and more beneficial to a plant that needs such an atmosphere.

Another first-rate solution to a too-dry apartment atmosphere is double potting. This, of course, is only an effective answer for potted plants. Double potting involves placing your plant pot inside a larger one and filling the space in between the two with moss. You can then water both the plant soil and the moss—continuing this as a regular procedure. The dewy moss will keep the humidity high in the plant's area and work quite well to defeat dryness. Make sure that the second, larger pot is of the non-porous variety.

Reverse your watering process every few weeks by watering from the bottom. This will do an excellent turn for the roots. You can keep your pot in a dish with pebbles to raise the pot above direct contact with the water. At the same time, the wet pebbles will keep the level of evaporation high and thus provide an increasing amount of moisture and humidity for the entire area of the plant.

Keep your plants near each other if this is possible. It will create an "outdoorsy" feeling that will be most beneficial for their growth. It will also keep the humidity high in the plant neighborhood. Granted, this may not be possible in a small apartment, due to restrictions of light or space—but if you have the room, it's a good idea.

If you live in the city, make it a practice to bathe the leaves of your plants once a month or so. This will keep them free of soot and dirt which is liable to clog the breathing pores of plants. Moist cheesecloth is perfect for this task.

Make it a point to keep your indoor garden away from drafts— prolonged exposure to low temperatures is not good for any of the growing things in your home.

If you're going away, you can count on your plants to do fairly well without water for at least a week. For any period longer than that, drape a plastic tent over the plant or plants—being careful not to place the tented area very close to direct sunlight. Of course, for many gardening projects more than a week's absence is just not going to work out too well. Use your own judgment—and if you do plan to be gone longer, perhaps a friend could be recruited as temporary caretaker of your indoor plot.

TROUBLESHOOTING

Often a plant will exhibit symptoms of illness that the gardener can't evaluate. We've listed the most common indoor plant ailments and their remedies for your convenience:

Plants Brown at Tips: This is usually a sign of overwatering— and with gardens that utilize chemical fertilizers it's a symptom of over-fertilization. The organic gardener needn't worry about the latter, you cannot administer an overdose of organic matter in the same way you might a chemical fertilizer.

Leaves Brown and Curled: Once again, this is liable to be the result of overwatering, or if you're quite sure that's not the case—

your room is too warm or the air is too dry. You can remedy this by making an attempt to lower general room temperature and/or increasing humidity by double potting or giving your plants some time in the bathroom.

Yellowed or Paling Leaves: This is likely to have something to do with the roots of your plant. If the roots seem to be winding around and around you probably need a bigger pot to accommodate them properly. Your plant might also be lacking in iron—so try to obtain an organic compound that will offer this to your deficient plant. Lastly, if the roots seem slightly rotten, it's a sign that you're overwatering.

Dropping Leaves: This is not a problem unless it is excessive and timely—that is, unless your plant is losing a good deal of leaves prematurely. This is a sign of too little humidity or too little fertilizer. You might want to try adding an increased amount of organic matter or varying the type you've been using already.

Sometimes you won't be able to pinpoint exactly what it is that is ailing your plant. If it seems to be just a lagging, depleted constitution—try a replanting if the plant seems sturdy enough for that. Or you might want to move the entire pot or container to a different part of the room. Often a change or variation in a plant's location will be quite beneficial.

For further information on the subject of indoor fruit gardening and indoor gardening in general consult the following informative sources: THE AFTER-DINNER GARDENING BOOK, Richard W. Langer, *Collier Books*; THE TROPICAL CROPS, O. W. Barrett, *Macmillan*; ARTIFICIAL LIGHT IN HORTICUL-TURE, A. E. Canham, *Centrex*; CITRUS FRUITS, H. H. Hume, *Macmillan*; THE MANGO, S. R. Gangolly, R. and D. Singh, *Council of Agricultural Research of New Delhi*; THE PINE-APPLE BOTANY, CULTIVATION AND UTILIZATION, J. L. Collins, *Interscience*; THE WORLD WAS MY GARDEN, D. Fairchild, *Charles Scribner's Sons*.

ECOLOGY

When we speak of gardening organically, we are contemplating, to a great degree, a new way of life. To comprehend and execute the principles of the organic approach to raising food is to attune yourself to something more than just a new method of gardening. Understanding the processes of nature, and how we all fit into her glorious and sundry schemes is the beginning of a new respect for our earth and, indeed, our entire environment.

As more and more people become seriously concerned about the progressive deterioration of our land due to mismanagement and misunderstandings, the Ecology Movement flourishes. What exactly is ecology? The dictionary defines it as the relation between living organisms and their environment. Beginning to comprehend exactly what these relationships are and the tremendous effect they have on everything that lives is the basis of the Ecology Movement itself.

Once we understand that there is no real way to deal with nature in a fragmented manner, we are well on our way to learning the importance of nature's totality, and the necessity for working with nature in this same, all-encompassing, interrelated fashion. When we realize that we cannot poison our soil without eventually poisoning ourselves, when we see that rampant destruction of insect life through chemicals disturbs a multitude of life processes—then we are beginning to grasp the interlocking quality of life upon this planet.

Organic gardening is based on a respect for the inextricably bound cycles of nature. It warns against poisoning the soil and plants, the water we drink and the air we breathe. Organic gardening sanctions only natural methods of working with the earth, and, in this way, it is truly an integral part of the Ecology Movement.

In many cases it is a lack of knowledge that permits people to contribute to the pollution of air, land and water that we see everywhere around us today. By availing yourself of information on the subject of organic living, you are arming yourself with the greatest tool any of us can hope for in this struggle to save our environment. And as you begin to use your knowledge in a practical way—by gardening, whether it be indoors or out—you have really become a participant in this crucial and highly honorable crusade.

When you make the decision to resist the seeming convenience of chemicals in your gardening project, you are insuring that the part of this planet that you have some "jurisdiction" over will be as pure and untainted as possible. And furthermore, by boycotting dangerous chemicals offered commercially you are helping to hasten the withdrawal of these deadly products from the market. There is a considerable movement on to ban the use and sale of DDT, but as long as people continue to purchase this product, it is unlikely that such a measure could be enforced. By refusing to use DDT in your garden, you are aligning yourself with organic gardeners and farmers the world over, and you have made a strong statement about your commitment to the need for sound and ecological living. It may seem insignificant to you that your tiny garden is chemical free—but if more people could make this decision, eventually the effects would be evident and highly beneficial on a large scale. Each of us can only cultivate our own garden —and if we can do it purely and ecologically, then we've done quite a lot.

When we begin to make use of the law of return in our gardening efforts, we are acting ecologically on several levels. We are giving the soil just exactly what it needs to stay strong and fertile, so that it may continue to produce nourishing and wholesome foods for our consumption. If we're working with land that is ailing, then by the use of organic materials we are rendering invaluable first aid to the land. Soil that is treated organically will maintain its all-important spongy structure covering and because of this will resist erosion. One of the major problems of our atmosphere today is the loss of the topsoil through erosion and other conditions which come about as a result of unhealthy soil. We can save our topsoil and in turn all the precious life that grows in and upon it by gardening organically. To save our soil is to save our entire planet—and this is not an exaggeration. Keep in mind that the soil is the very basis of all plant life. Without soil there would be no vegetation and without vegetation there could be no animal life. By strengthening this vital link in the chain of life we are giving a boost to all levels of living matter. It is indeed folly to believe that we can mistreat our soil, allow it to erode at an alarming rate, and not feel the effects ourselves. We can buoy up the resources of the soil and at the same time increase the benefits that plants and vegetables offer us.

Bodily health is certainly a major factor in ecology. We can't divorce the state of our own health from the health of the land we live on. Leaves and the green parts of plants are the basic factory of *all* food. The energy that we rely upon food to give us actually begins within the food factory of the vegetation that grows from the ground. This is the starting point for all food—and no matter how much it's been synthesized or chemicalized, there are basic ingredients that were obtained from the earth. When we garden organically we are supporting a "body ecology" movement. By insuring healthy plants through proper natural gardening methods we are guaranteeing ourselves quality food substances that will support us in the best way possible. The amount of energy we get from food depends upon the nutrient value of the plant and the soil that nurtured it. By keeping our gardens pure and free from poisons we are producing food that will be equally pure and wholesome. It is the return of organic substances to the soil that will provide the enrichment that will foster hardy and healthy plant growth. And so, organic gardening is a sure way to strengthen your body and give yourself the sanest and most natural kind of nourishment possible. When we speak of the Ecology Movement it is important to remember that human health is a very real factor in the scheme of interrelated life processes. If we are concerning ourselves with the health of our environment, then we must also take a serious look at the health of the people who dwell within it.

Many people today are purchasing tremendous supplies of drugstore-bought vitamins, but continuing to sustain themselves on a diet of synthetic, chemically treated foods. All the vitamins you could purchase at a store, however, can't compare with the natural nutrient value of the vitamins available from healthy, wholesome vegetables. Natural vitamins are provided for us by nature—but they are only present in sufficient amounts in well-balanced, organically rich soils. By striving to maintain this balance and make certain that the soil is amply supplied with com-

posted humus and other organic material, we are providing ourselves with a good slice of nature's wealth. Natural food vitamins
form an important part of the concept of preventive medicine.
That is, by seeing to it that we are properly nourished we can
avoid much illness and disease—and when we speak of this in
terms of millions of people, then indeed we are speaking of an
ecological precaution. And by way of assisting in overall good
health for all human beings, organic gardening bears strong relation to the ecology movement.

One of the greatest outcries today is against increasingly toxic
pollution of the air we breathe. With industry pouring poisonous
wastes into the air, it is all the more important that the home gardener not contribute to this unhealthy situation. To use chemical
sprays and additives in the soil is detrimental not only to the
ground, but to the air around us—the air we all take into our lungs.
To avoid use of such toxins in the air around our homes is to
contribute significantly to the measures now being taken to cut
down on air pollution. If you are fortunate enough to live in an
area which is not high in pollution to begin with, then certainly
it is your duty to keep the air as pure as you can. The use of organic
gardening practices will keep your "home air" from being pol-
luted and at the same time contribute to the healthiness of the
air by providing for the growth of healthy plants that will breathe
oxygen into the air. And perhaps this is one of the single most
beneficial aspects of gardening—the valuable addition of oxygen
to the air. Plants utilize the carbon dioxide that humans breathe
out as waste, and in turn they give off oxygen. This is one of the
most simple and exquisite exchanges we can observe in nature—
and it is to our benefit to encourage it as much as possible. So,
organic gardening serves a two-fold purpose in improving our
air: it avoids the pollution that chemical methods create and it
provides for increased oxygen content in the air. This is especially
important for the city and indoor gardeners, who are often likely
to feel that there is literally "no air to breathe." Plants growing in
your living room will make a significant contribution to the oxygen
availability and do much to cut down on the "stuffiness" of small
areas. On a larger scale, outdoor city gardening will put oxygen
into the air and at the same time utilize the overabundance of
carbon dioxide in the atmosphere. Surely the way to cleaner air

is not to spray the chlorinated hydrocarbons or organic phosphates that constitute chemical gardening substances into it. Such practices only increase the toxicity of the air and jeopardize the health of every living thing that must breathe that air. We do not receive all our nourishment from food—much of it comes from the vital elements in our air—and to make certain that this air is as pure and rich as possible, it is necessary to treat it kindly and wisely. In other words, don't put anything into the air that you wouldn't want in your lungs! It's that simple, and unfortunately, that frightening.

Organic gardening seeks to build up the resistance of plants to disease and insect pests by strengthening the soil and providing increased nutrition for the crops. Soil that is high in organic content will produce a strain of vegetables that is sturdy and vital; such plants needn't depend on chemical extinction of weeds and insects to survive. We have already seen how indiscriminate killing of plant life (weeds) and insects can disrupt the entire life cycle of soil and vegetation. This is because chemical campaigns of this sort cannot be selective. In other words, you can't compute a chemical pesticide to destroy only the harmful organisms. Instead you destroy thousands of helpful living things at the same time; and in destroying these organisms you lose the beneficial work they would have done for your garden. The soil is actually pulsating with millions of microorganisms, and many of them are crucial to overall soil and plant health. Ecologically speaking, it is necessary to realize the inherent interrelatedness of these organisms and work to perpetuate them, that they might serve the entire cycle of life to the fullest. It is not ecological to remove whole populations of organisms from the soil. Any attempt to destroy on a seemingly isolated basis will only result in havoc for all life processes. There is absolutely no way around this cause and effect chain of events in nature—and admitting this fact is what ecology is all about! Nature cannot be dealt with on a symptomatic level— if there is trouble in our vegetables then there is trouble in the soil. Organic gardening stresses dealing with the very core of any problems that arise—going to the source of each symptom of disease and coping with it at that point. This is successfully achieved by the strengthening of the soil with organic materials. To merely kill weeds or insects chemically offers no *positive* energy to the soil or the life upon it. To kill and extinguish is not the answer—and the more we go about this unthinkingly the more vicious the chemical cycle becomes. When we work symptomatically with nature we are causing tremendous imbalance and interruption of crucial inner workings—we are causing a loss of proportion and efficient functioning. One of the major concerns of ecology-minded people of today is to stress the need for a return to the balances of nature. Many facets of technology have contributed to the escalation of imbalance we see all around us today; organic gardening seeks to reinstate the necessary equilibrium of life.

Nature does not believe in waste. In fact, within the flow of nature there is no such thing. Anything that ceases to live, by the laws of nature, is returned to the earth, so that its valuable life elements may be used again in the growth process. It is only man's interference that has rendered the law of return inactive and created the problem of waste disposal. What we call garbage is actually organic material which could be returned to the soil as a benefit rather than a nuisance. When we interrupt the workings of the law of return, we find ourselves with a lot of superfluous material and nowhere to put it. The very basis of the compost method is to utilize these substances. In this way not only do they cease to accumulate and become a problem, but they are transformed into life-perpetuating treasures! The waste we're talking about does not simply take place in people's kitchens—it is an accepted part of big city industry as well. If individuals would begin to use their domestic wastes for organic farming compost work, and municipalities decided to think in terms of large-scale waste conversion, we would indeed be well on our way to a healthier environment.

City sewage and garbage can be turned into sludge—a highly useful compost material. At the present time, this country disposes of about 18 billion pounds of sewage solids per year. To use this vast amount of material in soil enrichment projects would not only go far in revitalizing our land, it would lessen the problem of water pollution and water-borne diseases that are serious issues today. Use of garbage and sludge would also cut down on the air pollution that results when these materials are incinerated or burned in town dumps. What we are talking about essentially is a recycling of wastes of all kinds. Organic gardening on a large

scale in this country would solve many serious problems at once—
indeed it is in many ways a panacea for an ailing environment.

You can be a practicing ecologist right in your own home. As
you begin to think and garden organically, you will see that many
things will fall into place. You'll have less garbage and more
oxygen—and healthier food to eat to boot! The ecological properties of the organic way of life are many—and just as the activities
of nature flow into one another—so the benefits of living ecologically overlap. You can become an integral part of this sane and
healthful system by embarking on your career as an organic gardener and an organic thinker!

FINAL WORDS

We have done our best to provide you with a well-rounded view into the workings of organic gardening. At the same time, we hope to have impressed upon you that the rewards of growing your own food, your own way, can be yours!

There are several pieces to the puzzle and we've attempted to fit them together in a way that is clear and comprehensive. If you can understand the basic concept behind organic living and the harm of the present synthetic approach to life, then you are no doubt anxious to start on the path to natural gardening and eating.

Keep in mind the need for patience and a good deal of persistence in your gardening attempts. Keep your pioneering spirit front and center at all times, if possible! You're going to be trying something you may have never dreamed possible, and if you're loyal to the belief that you can garden no matter who or where you are—you'll reap more than just a tasty vegetable crop.

You've become a part of an ever-growing movement. It's a movement of people determined to return to a healthier and more real way of life. Through organic gardening and eating, and increased ecological awareness, people such as yourself are striving to change our world for the better. The struggle for cleaner air, purer water, and chemical-free soil is a fight well worth the effort. So keep plugging—you're in excellent company!

FOOD VALUES

The following chapter is devoted to providing you with a ready reference guide to the make-up of almost every food you can think of. Designed to assist you in making some gardening choices, we're certain that this section will be valuable in helping you tailor your entire diet to include the food you feel will be most beneficial to you. For each food we've pointed out the vitamins, minerals, etc., they are likely to contain and sometimes what they do not contain, so that you may, if you like, decide to eliminate some of these things from your daily eating regime. We've boiled down a vast amount of information into straight facts (and figures) and decided to present it to you in a terse, no-nonsense fashion that enables us to bring it to you in its most complete form.

FOOD VALUES

SPECIAL NOTE

The information presented in this chapter will aid the reader in gathering knowledge of the particular nutrients found in foods common to the American diet. No reference is intended to mislead the reader into thinking that any one food or any combination of foods, will *cure* or reduce any ailment. We do not believe in the *cure* or *prescriptions* which other authors in the Health and Nutrition field have tried to force upon the reader. When we state, then, that Papaya will aid digestion it does not mean that it will also cure indigestion. Please accept the following information as a guide only. It is not a prescription.

To compile the information for this chapter required the examination of more than fifty books. The primary reference books on this subject are "The Complete Book of Food and Nutrition,"
J. J. Rodale & Staff, *Rodale Books Inc.* "Dictionary of Foods" Hauser & Berg, *Beneficial Books.* "About Nuts and Dried Fruit," P. E. Norris, *Thorsons Publishers Ltd.* "Vitamins: what they are and how they can benefit you," H. Borsook, Ph.D., M.D., *Pyramid Books.*

The average person takes it for granted that whatever he eats has sufficient nutritional value to get him through the day. This is rather evident due to the enormous quantity of "Quick Service" foods consumed every day by millions of people. Unfortunately these consumers are unaware that hot dogs, pizza, french fries, and colas have little or no nutritional value and are, in addition, unhealthy.

Lack of basic knowledge concerning food values is quite common throughout America. Few housewives know what vitamins, minerals, fats, acids or protein are in any given food. Much less how various foods should be combined. The catchall phrase, "a balanced diet," means little when people do not know what they are balancing, or have been deceived about what they believe they are consuming.

If people were made aware of the complete lack of food values in various popular foods, it might be assumed that they would make efforts to correct their eating habits. Virtually all foods obtainable through normal supermarkets and groceries are adulterated to one degree or another through the use of preservatives, coloration, pesticide residue, and a variety of other poisons (see chapter on POISONS). We have used organic sources in arriving at the following food definitions except where specifically stated otherwise.

The following chapter deals with food values and will be a useful reference and guide to understanding nutritional values of all foods common, and sometimes not common, to the American diet.

NOTE

To avoid confusion in the use of the terms *fats* and *oils*, we might begin by saying that the fat content of nuts is very high—in some instances well over 50 percent.

There is no food over which human digestive capacity varies so much as fat, but fat in one form or another is necessary.

All fats consist of three elements; carbon, hydrogen, and oxygen. They are the most concentrated forms of fuel and give twice as much heat as either proteins or carbohydrates. When digested they are absorbed partly into the small intestine and partly into the lymph, and ultimately into the bloodstream. Fat is stored in depositories under the skin; in the tissues of most organs; in the mesentery and around the kidneys; and when these fat storage deposits are replete, fat eventually passes into the liver.

You can cut down your intake of fats but there is a definite need for some fat to be consumed in your diet. Fats in the form of phospholipids play an important role in nourishing the brain, and a class of compounds derived from fats known as sterols are essential to life.

As far as we can ascertain fats and oils from nuts do not contain vitamin D, though margarines made from nut oils can be fortified to contain this very important vitamin and are in this respect just as beneficial as butter.

It was once thought that nut butters and oils were devoid of vitamin A, but through modern research scientists have found that some nut oils contain substantial quantities. Nuts also contain a considerable quantity of basic amino-acids; butter made from nuts has a high biological value. They not only help to maintain life and growth, but if fed to nursing and pregnant mothers improve the quality of their milk.

Unless nuts are ground by the teeth and broken down, the digestive juices can work only imperfectly on them and they pass undigested into the alimentary canal. The digestibility of nuts increases by as much as 10 percent when turned into butter. Unfortunately, most commercial nut butters are made from roasted nuts which have been heavily salted, and the excessive heat under which they are treated develops free fatty acids. Nuts should be dehydrated at moderate temperatures or gently dried to make them crisp. If roasted at high temperatures, the B complex vitamins are destroyed. In addition, salting nuts, far from improving their digestive qualities, interferes with the process of digestion.

Nuts are one of the most concentrated foods and have as high a nutritional value as any animal product with the exception of cheese. The fat in all nut butters is more easily digested than butter. Animal fats are free fats and will not mix with water whereas nut fats do so and form an easily digestible emulsion.

If you have had any difficulty digesting nut butters, eat them in combination with vegetables. In order to digest fats, there must also be present organic sodium in your food. Sodium is the principal alkaline element in the process of saponification which takes place when the pancreatic juice, bile and intestinal juices come into contact with fats. The organic sodium found in fresh vegetables and fruits will accomplish this.

When eating nuts, leave the skins on, because the vitamins are usually found in this covering. These vitamins are destroyed when the nuts are salted and roasted when the fats are split into glycerine and free fatty acids.

NOTE

Since dried fruits are not mentioned in the list of food values, special notice to their nutritional qualities will be given below.

Dried fruits are important in their own right and not merely as a substitute for fresh fruits. The vitamin C in dried fruits is not as high as that of fresh fruits, but drying concentrates the sugar, and makes it an infinitely greater source of energy.

In the U.S.A. the law stipulates that there shall not be more than 350 parts of sulphur dioxide per million, but this law is often broken and much dried fruit from this country has as much as 1,000 parts per million.

Peaches, pears, nectarines and apples are prepared for dehydration by having the stones or cores removed and are then placed in a "sulphur box." This is done so that 1. Fruit may appear as transparent as possible; 2. to reveal decayed portions overlooked in trimming; 3. to prevent fermentation and decay while drying; 4. to kill flies and other insects which leave behind larvae, which may develop after storing; 5. to make the texture of the fruit porous and so facilitate drying.

Sulphured fruit generally looks more yellow and attractive which adds to its sales potential.

Sulphured fruit contains more water than sun dried fruit and weight for weight is of less value. In many packing houses the fruit is reprocessed; that is, washed and resulphured, increasing the sulphur content to 3,000 parts per million or more, which makes the fruit unhealthy.

Sulphur adds nothing to the flavor of fruit or its quality. Sulphurous acid interferes with the functions of the kidneys, which have to work harder to remove it from the system; and in many instances, sulphur has caused the retardation of the red blood corpuscles.

There is no need for sulphur dioxide to be used in dehydration and many manufacturers have stopped using it.

The most important value of dried fruits lies in their carbohydrate content and mineral matter. In order to maintain optimum health one must have fats for fuel and energy; minerals to supply bone, blood and fluid substances; vitamins which have many beneficial functions and carbohydrates which also supply heat and energy.

Carbohydrates form the bulk of the food most people eat: bread, potatoes, all vegetables, fruits and some nuts.

There are three types of carbohydrates: 1. sugars, 2. starch, 3. cellulose, gums and pectins. Carbohydrates are composed of hydrogen, oxygen and carbon. Hydrogen and oxygen are present in the same proportion as water: two to one. The fuel value of carbohydrates is only half that of fats because the hydrogen in carbohydrates is already combined with oxygen in the form of water and so it cannot be burned as it is in fats.

Starch ends up as sugar in the body. Some forms of sugar are assimilated straight into the bloodstream while others, in a more complex chemical form, have to be broken down before this can be accomplished.

After passing into the bloodstream they emerge from the intestines and pass to the liver to be stored in the form of glycogen. As much glucose as is needed for immediate use passes on and is transformed into glycogen in the muscles, which use up glycogen as they work, which causes a rather constant need for it.

The liver is the main storehouse of glycogen. The amount of carbohydrate needed for immediate use goes into the muscles and the remainder is stored in the liver. Even though the sugar content of dried fruit is far greater than fresh fruits, you should eat both because fruits contain vitamins, juices and acids. Dried fruits are the richest of all in minerals and only sea foods, such as clams, lobsters and oysters can compare with them in this respect. Minerals found in dried fruits are potassium, sodium, calcium, magnesium, iron, sulphur, silicon, and chlorine.

CHEMICAL COMPOSITION OF NUTS AND DRIED FRUITS

NUTS	Water	Protein	Carbo-hydrate	Fat	Minerals
Acorns	4.10	8.10	48.00	37.40	2.40
Almonds	4.90	21.40	16.80	54.40	2.50
Beechnuts	9.90	21.70	19.20	42.50	3.86
Brazil Nuts	4.70	17.40	5.70	65.00	3.30
Butternuts	4.50	27.90	3.40	61.20	3.00
Candlenuts	5.90	21.40	4.90	61.70	3.30
Chestnuts (dried)	5.90	10.70	74.20	7.00	2.20
Chufa	2.20	3.50	60.70	31.60	2.00
Coconut	14.10	5.70	27.90	50.60	1.70
Filberts	5.40	16.50	11.70	64.00	2.40
Hickory Nuts	3.70	15.40	11.40	67.40	2.10
Paradise Nuts	2.30	22.20	10.20	62.60	2.70
Pecans	3.40	12.10	8.50	70.70	1.60
Pignons	3.40	14.60	17.30	61.90	2.90
Pignolias	6.20	33.90	7.90	48.20	3.80

NUTS	Water	Protein	Carbo-hydrate	Fat	Minerals
Pistachios	4.20	22.60	15.60	54.56	3.10
Black Walnuts	2.50	27.60	11.70	56.30	1.90
English Walnuts	2.50	18.40	13.00	64.40	1.70
Water Chestnuts	12.30	4.00	50.00	1.20	1.77
Peanuts	7.40	29.80	14.70	43.50	2.25
Peanut Butter	2.10	29.30	17.10	46.50	2.20
Almond Butter	2.20	21.70	11.60	61.50	3.00
DRIED FRUITS					
Apples	26.10	1.60	62.00	2.20	2.00
Apricots	29.40	4.70	62.50	1.00	2.40
Pears	16.50	2.80	66.00	5.40	2.40
Peaches	20.00	3.15	50.00	.45	2.15
Prunes	22.30	2.10	71.20	—	2.30
Raisins	14.60	2.60	73.60	3.30	3.40
Currants	17.20	2.40	74.20	1.70	4.50

In order to clarify some of the values mentioned in this book we have prepared the following chart on the vitamin contents of common foods. This chart was originally printed in the best seller "Vitamins: What they are and how they can benefit you," by the nationally acclaimed nutritionist Henry Borsook, Ph.D., M.D., published by *Pyramid Books*. We reviewed dozens of food value charts for this chapter and found the following one by Dr. Borsook to be the most accurate and comprehensive of any in its field. In addition, whenever we mention that a particular food is high, medium or low in a certain nutrient, you can compare it with this chart to get a more specific and detailed analysis of the food in question.

THE VITAMIN CONTENTS OF COMMON FOODS

Unless otherwise indicated, the quantities given below are those in the dry or fresh state before cooking.

No values for vitamin D are given because except in milk, butter, and eggs, the amounts in all foods are too small to be significant.

The data is as yet too limited in research findings, or the daily human requirement is inadequately determined, to make it useful to include in this table values of vitamins B6, nicotinic acid, pantothenic acid, E and K. However, for a detailed analysis of each of these vitamins, read the chapter on VITAMINS.

Where no vitamin values are given in this table it is because they have not yet been determined. In any case, the amount present probably is insignificant for ordinary nutritional purposes, except in the case of vitamin B2.

UNIT EQUIVALENTS:

Vitamin A
1 International unit = 2 Sherman Units
= 0.6 microgram (gamma, y) of B carotene

Vitamin B1 (Thiamine Chloride)
1 International unit = 3 micrograms (gamma, y)
= 0.003 milligram
= 2 Sherman units

Vitamin C (Ascorbic Acid, Cevitamin Acid)
1 milligram = 20 International units
= 2 Sherman units

Vitamin B2 (Riboflavin)
1 milligram = 333 Sherman-Bourquin units
= 1000 micrograms (gamma, y)

	MEASURE		Vita-min A I.U	Vita-min B1 I.U	Vita-min C Mill.	Vita-min B2 Mill.
BREADS & CEREALS	Ordinary	Ounces				
Barley, whole grain	1 tblsp.	½	0	3.3	—	0.001
" , pearled	1 tblsp.	½	0	0	—	
Biscuits, baking powder	1 biscuit		19	3	—	
Bread, Boston brown	1 slice 3″ diam. ⅜″ thick	¾	55	13	—	
" , white, made with milk	1 slice 3″x3½″x½″	¾	10	3.8	—	0.0121
" , white, made with water	1 slice 3″x3½″x½″	¾	10	3.7	—	0.006
" , 100% wheat	1 slice 3″x3″x⅜″	1	70	27	—	
" , rye	1 slice 3″x3″x⅜″	1		14	—	
Corn, whole grain yellow	1 cup	5	1,200	72	11	0.05
" , flakes, cereal	1 cup	1½		trace	0	
" , meal, white	1 cup	5	0	143		
" , meal, yellow	1 cup	5	1200	110		
Crackers, Graham	1 cracker		26	8	0	
Custard, baked	¾ cup	3	650	21	1	
Flour, rye	1 cup	5		30–70		
" , 100% whole, unbleached	1 cup	4	500	180	0	
" , white, bleached	1 cup	4	150	33	0	
" , white, pastry	1 cup	4	130	19	0	
" , white, plus germ	1 cup	4		49	0	
Griddlecakes	1 medium	2	200	11	0.5	
Muffins, plain w. egg	1 muffin		135	9	0	
" , plain, without egg	1 muffin		48	8	0	
" , bran, with egg	1 muffin		260	50	0	
" , bran, without egg	1 muffin		170	52	0	

	MEASURE Ordinary	Ounces	Vitamin A I.U	Vitamin B1 I.U	Vitamin C Mill.	Vitamin B2 Mill.
BREADS & CEREALS						
Oatmeal, whole grain	½ cup	1¾	0	165		
" , quick cooking	½ cup	1¾	0	130	0	
Oats, rolled, packaged	½ cup	1¾	0	121	0	
" , rolled, cooked	½ cup	1¾	0	121	0	
Rice, brown	2 tblsp.	1	17	15		0.02
" , polished	2 tblsp.	1	0	0		0.022
Rolls	1 holl	1½	74	9	0	
Rye, whole grain	1 cup	5		210	0	
Wheat, whole grain	1 tblsp.	½	50	23	4	0.02
" , whole grain, cooked	1 tblsp.	½	50	23		
" , bran	1 tblsp.	½	85	28	0	
" , farina, light	1 tblsp.	½	0	0.2	0	0
" , germ	1 tblsp.	½	90	80	0	0.10
" , puffed	½ cup	½		0	0	
" , semolina	1 tblsp.	½	30	7		
" , shredded	1 biscuit	1	4	20	0	0.10
Wheat, stone-ground	1 tblsp.	½	40	22	3	0.015
DAIRY PRODUCTS						
Butter	1 square	½	315	0	trace	0.001
Buttermilk	1 large glass	8	0	35	2	0.310
Cheese, Am. Cheddar	1" cube	¾	420	3	0	0.12
" , Camembert	1" cube	¾	750		0	
" , cottage, skim	1 tblsp.	¾	70		0	0.068
" , creamed, soft	1 tblsp.	½	310	0.6	0	0.017
" , creamed, full	piece 2"x1"	1	500		0	0.02
" , Edam	1" cube	¾	300		0	
" , pimento (kraft)	1" cube	¾	500		0	
" , Roquefort	1" cube	¾	850		0	
" , Swiss (kraft)	piece 4"x4"	¾	440		0	
Cream, 20% fat	1 tblsp.	3/5	64	2	trace	
Cream, 40% fat	1 tblsp.	3/5	132	1.6	trace	
Eggs, whole	1 egg	1½	900	19	0	0.25
" , white	1 white	9/10	0	trace	0	0.14
" , yolk	1 yolk	6/10	900	19	0	0.11
" , soft-boiled	1 egg	6/10	900	19	0	
" , hard-boiled	1 egg	6/10	900	19	0	
Milk, whole, fresh, raw	1 quart	32	1,400	150	19	1.20
Milk, whole, fresh, raw	1 glass	6	260	28	3.5	0.23
" , whole, fresh, pasteur	1 glass	6	260	22	3.0	0.23
" , dried, reconstituted	1 tumbler	6	260	20	2.0	0.20
" , evaporated	½ cup	4	190	15	3.5	0.35
" , condensed, sweetened	1 tblsp.	¾	45	10		
" , skim, fresh	1 glass	6	trace	25	4.0	0.17
" , skim, dried powder	1 tblsp.	¼	0	8	0.7	0.09
" , shake, ice cream		12	0	37	4	
MEAT & FISH						
Bacon, fried	5 slices	½		5	0	
Beef, lean, top round	¼ lb.	4	40	45	2.2	0.06
Chicken, light meat	¼ lb.	4		30	0	0.029
Chicken, dark meat	¼ lb.	4		42	0	0.078
Chicken liver	⅛ lb.	2	17,000	50	11	2.0
Cod, steak, fresh	¼ lb.	4	2	34	0	0.11
Crab	¼ lb.	4	2,200	45	5	0.40
Halibut, muscle	¼ lb.	4		32		0.21
Ham, smoked, lean	¼ lb.	4		540		
Herring, whole	¼ lb.	4	1,700	20		0.12
Kidney, beef or calf	¼ lb.	4	450	105	12	1.6
Lamb, chop, lean	¼ lb.	4	trace	90	2	

	MEASURE		Vita-min A	Vita-min B1	Vita-min C	Vita-min B2
	Ordinary	Ounces	I.U	I.U	Mill.	Mill.
MEAT & FISH						
Liver, beef, fresh	¼ lb.	4	46,000	100	34	3.4
Mackerel	¼ lb.	4		34		
Mutton, lean	¼ lb.	4		68		
Oysters, raw	⅓ cup	3½	420	75	3	0.46
Pork chop, lean	¼ lb.	4	0	515	2	
Pork, loin, lean	¼ lb.	4	0	515	2	0.28
Prawns, boiled	¼ lb.	4	1,100	20	0	0.11
Salmon, fresh, canned	¼ lb.	4	340	trace		0.27
Sardines, canned in oil	⅛ lb.	2	200	17		
Sweetbreads, fresh	¼ lb.	4		120		
Tongue, beef or sheep	¼ lb.	4		35		
Trout, fresh-water	¼ lb.	4		33		
Veal, muscle, cooked	¼ lb.	4	40	45	2	0.17
Whiting, Atlantic	¼ lb.	4	400			
FRUITS						
Apples, raw, average	1 medium	6	50	8	10	0.06
" , applesauce	½ cup	4½			9	
" , Juice	1 tumbler	6			5	
Apricots, fresh	1 medium	1⅓	1,000	3.5	2	0.04
" , dried	4 halves	3/5	540	8	0	0.009
" , dried, sulphured	4 halves	3/5	540	3	2	0.005
Avocado, California	½ avocado	4	64	40	25	0.19
Bananas	1 medium	5½	230	28	7	0.07
Blackberries	1 cup	5¾	230	10	15	
Blueberries	1 cup	5¼	150	21	8	0.021
Cantaloupe	½ mellon	13½	770	73	60	0.25
Cherries, fresh, bing	10 cherries	2⅓	75	12	13	
Cranberries, fresh	1 cup	3¾	20	0	20	0
Cranberries, Juice	½ cup	2			7	0
Currant, black fresh	1 cup	3¾		11	220	
Currant, red, fresh	1 cup	3¾		16	55	
Dates, fresh	4 dates	1½	35	12	0	0.012
Dates, dried	4 dates	1½	15	8.5	0	0
Figs, fresh	3 small	4	90	30	2.3	0.06
Figs, dried	3 small	2-1/5	32	30	0	
Gooseberries, canned	1 cup	4	330	56	34	
Grapes, white, seedless	½ bunch	4	22	11	2.2	0
Guava	1 guava	¾	22	2	30	0.006
Lemon	1 large	3-3/5	0		55	
Lemon, Juice	2 tblsp.	1	0		15	0.001
Lime juice, fresh	2 tblsp.	1	8		10	
Mangoes	1 mango	3	850	17	25	0.05
Olives, green	6 olives	2	85	trace	0	
Olives, mission	6 olives	2	28	trace	0	
Orange, pulp	1 large	9½	80–270	95	80	0.24
Orange, Juice, fresh	1 tumbler	6	540	59	98	0.012
Orange, Juice Canned	1 tumbler	6	540	59	72	
Papayas	1 papaya	3	1,800	7	46	0.015
Peaches, fresh	1 medium	4	900	28	3.4	0.008
Peaches, canned	2 halves	3	900	22	1.7	
Pears, fresh, Bartlett	1 pear	3-1/5	6	5	3	0.015
Pineapple, fresh	1 slice ⅜"	1⅜	90	13	2.6	0.006
Pineapple, Canned	1 slice ⅜"	1⅜	90	7	2.3	
Pineapple, Juice, canned	1 tumbler	6	90	35	12	
Plums, fresh	2 plums	3-4/5	210	17	5.5	0.027
Plums, canned	2 plums	3-4/5	200	15	5	
Pomegranate	1 pomegranate	6	0		20	
Prunes	4 prunes	1-2/5	350	24	trace	0.3

| | MEASURE | | Vita-min A I.U | Vita-min B1 I.U | Vita-min C Mill. | Vita-min B2 Mill. |
	Ordinary	Ounces				
FRUITS						
Quince	1 quince	4		11		
Raisins	¼ cup	1	28	21	0	0.04
Raspberries	1 cup	4-3/5	270	trace	33	
Strawberries	½ box	6	340	trace	93	
Tangerines	1 tangerine	2	280	22	28	0.022
Watermelon	1 slice 6"¾	11	0	25	15	0.18
VEGETABLES						
Artichokes	1 artichoke	8½	250	70	15	0.025
Asparagus, green	2 stalks	1⅓	130	22	19	
Beans, baked	1 cup	4	15	150	0	
" , kidney, fresh	½ cup	3	27	63	0	0.025
" , navy, dried	½ cup	3½	27	128	0	
" , lima, fresh	½ cup	3	0	100	13	0.25
" , string or snap	½ cup	3½	600	25	8	0.08
" , string, canned	½ cup	3½	600	11	2	
" , runner, green	½ cup	3½	700	40	5	
" , soy black	½ cup	3½	900	100	46	
" , soy, white	½ cup	3½	140	250		
" , wax, butter, yellow	½ cup	3½	410	30	16	
Beets, Root	½ cup diced	5	70	24	5	
Broccoli, fresh	1 cup	4	4,000	38	80	
Brussels sprouts	6 sprouts	4	1,100	65	100	
Cabbage, white, raw	1 cup	3⅓	800	25	25	0.04
Cabbage, white, cooked	1 cup	3½	50	20	12	
Carrots, raw	1 cup diced	4¾	2,700	32	8	0.03
Carrots, cooked	1 cup diced	4¾	2,700	32	5	
Carrots, canned, strained	1 cup diced	4¾	2,700	13	3	
Cauliflower, raw	¼ head	3	49	46	33	0.065
Cauliflower, cooked	¼ head	3		35	33	
Celery, green stems	2 stalks 7"	1½	320		2	
Celery, blanched stems	2 stalks 7"	1½	2	5	2	
Chard (beet tops) raw	½ cup	5-1/5	12,000	20	43	
Chard beet cooked	½ cup	5-1/5	12,000	15	28	
Chicory (escarole) raw	4 leaves	3½	9,500		5	0.022
Chives	1 teasp.	1/5			2	
Collards, fresh raw	½ cup	3-3/5	2,200	68	50	0.3
Collards, cooked	½ cup	3-3/5	2,200	50	22	
Corn, yellow, whole	¼ cup	1-4/5	250	25	5	0.018
Corn, sweet, canned	½ cup	3½	120	30	11	
Cucumbers, raw	½ of 10"	6½	0–20	55	18	0.003
Dandelion greens	1 cup	3½	12,500			
Eggplant	3 slices 4"	6⅓	70	43	8	0.10
Endive	3 stalks 6"	7⅓	4,200	70	20	
Garlic	1 clove	¼	0		2	
Kale, raw	1 cup	3-2/5	9,000	59	140	0.50
Kale, cooked	1 cup	3-2/5	9,000	45	28	
Kohlrabi	1 cup ½"	5			39	
Leek	1 leek 7"	1	17	8	7	
Lentils	2 tblsp.	1	25	14		0.022
Lettuce, Romaine	2 leaves 9"	¾	1,300	4	1	0.03
Lettuce, Iceberg, head	¼ L. Head	2½	150	20	4.5	0.05
Mustard greens	½ cup	2⅓	100	30		
Okra	5 pods	2	85	23	6	0.26
Onions, fresh	1 medium	2	6,200			
Onions, fresh, stewed	1 medium	2	0	10		
Parsley	4 stems	½	1,000		11	
Parsnips	17"x2"	6¾	380	66	38	
Peas, fresh	½ cup	2½	530	85	20	0.026

	MEASURE		Vita-min A	Vita-min B1	Vita-min C	Vita-min B2
	Ordinary	Ounces	I.U	I.U	Mill.	Mill.
VEGETABLES						
Peppers, green	1 pepper	2½	600	69	0.12	
Peppers, Red	1 pepper	2½	2,200		135	
Potatoes, yellow, sweet	½ medium	3-3/5	8,000	31	20	0.02
Potatoes, yellow, sweet, cooked	½ medium	3-3/5	4,000	25	10	
Potatoes, white	1 medium	5⅓	60	93	30	
Potatoes, white cooked	1 medium	5⅓	40	74	30	
Rutabagas	1 cup	5	10	28	22	0.13
Sauerkraut, fresh	1 cup	5		trace	16	
Sauerkraut, cooked	1 cup	5			3	
Sauerkraut, juice	2 tblsp.	1			4	
Spinach, raw	½ cup chop	3-3/5	14,000	37	53	0.065
Spinach, cooked	½ cup chop	3-3/5	14,000	30	26	
Squash, winter, Hubbard	½ cup	4	2,900	21	3.5	
Squash, summer winter	½ cup	4	170	16	3.5	
Tapioca	2 tblsp.	1	0	0	0	0
Tomatoes	1 medium	7	6,000	52	50	0.10
Tomatoes, canned	½ cup	4¼	3,600	25	24	
Tomatoes, juice, canned	1 tumbler	6	5,300	25	25	
Turnip greens	½ cup	3	6,000	38	42	0.25
Turnip, cooked	½ cup	3	6,000	25	10	
Watercress	½ bunch	1½	2,000	28	22	
NUTS						
Almonds	10 nuts	3/5	0	7	0.5	
Cashews	5 nuts	½	19			0.028
Chestnuts	2 nuts	½		12	5	
Cocoanut, shredded	1 tblsp.	⅓	0	1	0.5	0.014
Cocoanut, milk, fresh	1 cup	8	0	0	4	
Hazel	10 nuts	½		28	2	
Peanuts, whole, spanish	⅓ cup	1½	17	146	4	0.3
Peanuts, shelled	⅓ cup	1½	17	146	4	0.3
Peanuts, roasted	⅓ cup	1½		33	0	
Peanut butter	2 tblsp.	1½	17	125		
Pecans	12 nuts	½	14			
Pistachio	15 nuts	½	14			
Walnuts	12 nuts	½	190	16	3.5	
MISCELLANEOUS						
Corn oil			0	0	0	0
Cottonseed oil			0	0	0	0
Ice cream, made with skim milk		8	0	15	3	
Lard			0	0	0	0
Mayonnaise	1 tblsp.		30	0.5	0	
Pie, apple	1 serving		90	14	0	
Pie, blueberry	1 serving		62	9	3	
Pie, chocolate	1 serving		440	10	0.5	
Soup, black bean	1 cup	4	70	11	0	
Soup, split pea	1 cup	4	145	18	0	
Soup, tomato	1 cup	4	1,480	10	9	
Soup, vegetable	1 cup	4	276	9	4	
Sugar			0	0	0	0
Tea			0	0	0	0
Coffee			0	0	0	0
Cocoa	2 tblsp.	½	0	4	0	0
Beer	1 glass	6	0	4	0	0
Yeast, baker's compress	1 cake	2	0	100	1	1.4
Yeast, baker's dried	1 cake	½	0	70		0.5
Yeast, brewer's, fresh	1 cake	2	100	220		0.85
Yeast, brewer's, dried	1 cake	½		8–500		0.4

VITAL ELEMENTS IN COMMON FOODS

CALCIUM
cheese (Am.)
milk (whole)
cheese (Cot.)
milk (Butt.)
cauliflower
broccoli
endive
celery
beans
rutabagas
spinach
turnips
carrots
molasses
oysters
string beans
cabbage
lettuce
eggs
nuts
fish
almonds
citrus fruit
maple syrup
fruits dried
beans dried

SULPHUR
Watercress
asparagus
cabbage
garlic
grape
onion
beans
bran
bread
brussels S.
cauliflower
cheese
clams
eggs
fish
meat lean
nuts
oats
oysters
peas
chard
turnips
wheat

POTASSIUM
cabbage
cocoanut
figs
tomatoes
apricots
peaches
onions
lima beans
pineapple
milk
prunes
pears
string beans
egg plant
watercress
celery
cauliflower
raisins
potato
citrus fruit
parsley
dates
rhubarb
carrots
spinach

SODIUM
wheat bread
rye bread
buttermilk
cream cheese
codfish
halibut
mackerel
salmon
bananas
celery
dandelions
lettuce
spinach
sweet potato
milk
am. cheese
beet
watercress

MAGNESIUM
cocoa
chocolate
almonds
cashews
peanuts
lima beans
whole wheat
brown rice
oatmeal
dates
raisins
chard
spinach

CARBON
potatoes
brown sugar
whole wheat
shred wheat
honey

PHOSPHORUS
lima beans
cheese Am.
oatmeal
veal
beef liver
beef lean
fish
eggs
spinach
buttermilk
milk (Skim)
milk (whole)
almonds
grapes
lentils
pecans
rice brown
walnuts
whole wheat
brussels S.
corn
dandelion
lobster
peas
soybeans

IRON
beans
beef liver
egg yolk
peas
wheat
oatmeal
beef lean
prunes
spinach
parsley
kale
cheese
potato
chard
watercress
oysters
dates
raisins
beets
figs
oranges
mushrooms
turnips
tomatoes
banana
carrots

IODINE
lobster
clams
oysters
shrimp
blue fish
mackerel
haddock
cod
halibut
scallops
salmon
squash
radishes
asparagus
lettuce
milk
cabbage
cucumber
string beans
spinach
beets
potato
kelp
sea lettuce

HYDROGEN
vegetables
fruits
milk
water

NITROGEN
cheese
fish
lima beans
peas
lentils
mushrooms
cheese
meats
nuts

OXYGEN
in all food
and water

MANGANESE
pineapple
wheat
beans navy
blueberries
tangerines
walnuts
kidney beans
beets
liver, calves
lima beans
beets
wheat bran
banana

gooseberries
spinach
peaches, dry
blackberries
green olives
apple
apricot
beet greens
cabbage
carrot
celery
lettuce
watercress

FLUORINE		SILICON		CHLORINE	
cauliflower	sauerkrout	asparagus	strawberries	oysters	dates
cod liver	sea food	spinach	cherries	cheese	dandelion
oil	rye bread	lettuce	apples	lettuce	cocoanut
goat's milk	cabbage	tomatoes	celery	spinach	carrot
egg yolk	whole	barley	beets	whey	tomatoes
cheese	grains	figs	parsnips	cabbage	bananas
brussel	spinach	berries	black figs	parsnips	onions
sprout	watercress	oatmeal	radishes	beets	pineapple
milk	meat broth	bran	chard	turnips	eggs
garlic	beets	grapes		milk	grapes
				watercress	oranges
				fish	lemons
				celery	wholewheat
				cottage cheese	

Vegetable and Fruit Juice Food Values

Juices	Vitamins	Mineral	Good For
Apple	A, B2, C	Magn., Potas., Sodi., Calci., Chlo., Iron	Bleeding gums, digestion cleansing the intestines.
Beet	A, B1, C	Sodi., Calci., Potas.	Liver, gall bladder, building red corpuscles.
Cabbage	A, B, K, E, U	Chlo., Calci., Sodi., Iron.	Muscle pains, bleeding, gall bladder.
Carrot	A, B, C, E	Potas., Iron, Magn., Sodi., Silicon, Sulph., Phosphorus	Anemia, stomach and digestive ailments.
Cucumber	B, B1, C	Potas., Iron, Magn., Chlo.	Constipation, Kidney trouble, cleans blood.
Celery	A, B, C, D1	Sodi., Magn., Iron, Calci.	Rheumatism, nervousness, skin problems.
Grape	A, C	Iron, Potas.	Anaemia, body energy.
Lemon	C, P	Sulph., Potas.	Common cold.
Orange	A, B, C	Calci., Phos.,	Anaemia, loss of energy
Potato	A, B2, C	Potas., Sulph.	Stomach acid, skin problems.
Radish	A, C, D	Chlo., Phos.	Intestinal troubles.
Strawberries	C, D	Iron, Fruit Sugar	Anaemia, constipation, skin eruption, glands, nerves.
Tomato	B, C, D, K	Magn., Iron, Chlo., Sulph., Sodi., Iodine	Liver, gall bladder,. nerves.

NOTE: Magn.-magnesium. Sulph,-sulphur. Chlo.-chlorine. Potas.-Potassium. Phos.-phosphorus. Calci.-calcium. Sodi.-sodium.

Agar-Agar

A species of sea grass, commonly known as vegetable gelatin. It consists of carbohydrates and traces of iodine.

Ale

Fermented extracts obtained from germinated, starchy seeds; an infusion of malt with the addition of hops, which results in a mildly stimulating alcoholic beverage. It contains excessive amounts of phosphorus acid and only a small amount of vitamin C.

Almond

A nut. The kernel contains a large percentage of protein, a good deal of fat and only small amounts of vitamins A and B; having an acid excess. Almonds are very difficult to digest if not properly chewed.

Apple

A tree fruit. The wild or crab apple is more preferable from the nutrition standpoint than the over-cultivated variety. Apples aid the digestion and when eaten before a meal will help overcome constipation due to their fruit-acid salts which stimulate the digestive processes and their semi-cellulose content which softens the partly digested food in the large intestine. They contain a slight alkali excess and only slight traces of the water-soluble vitamins B and C. Fresh, uncooked apple juice also aids digestion.

Apples contain 84 percent water, 0.4 percent protein, 0.5 percent fat, 13 percent carbohydrates and 0.5 percent mineral matter. An apple may be eaten raw or cooked; grated in salad or as apple sauce. While they do not combine well with potatoes, bread and pastry, they can be eaten with yogurt and several other foods (see chapter on Food Combining). Apple pectin is a natural germicide. It not only acts as a killer of organisms but also stimulates the growth of healthy tissue. Pectin is also useful as a gelatin in preserves and jellies.

Apricot

The ripe fruit of the apricot tree is rich in potassium and iron and contains a high alkali excess. Apricots are extremely rich in vitamin A. One medium apricot (1¼ oz) contains 1,000 I.U.

of A. Four dried halves (3/5 oz) contain 540 I.U. of vitamin A. In cases of liver or digestive disturbances apricots should be cooked or soaked well before eating.

Artichoke

This bud of the flowering head of a large thistle is very rich in potassium, sodium and calcium; being a bud vegetable, over-rich in phosphorus and sulphur acids and thus causing an acid excess; containing small quantities of water-soluble vitamins, and a good amount of the fat-soluble vitamin A; possessing an oxidizing ferment (tyrosine) and an insulin having the same properties. This bud is rich in tannic acid and, therefore, is slightly constipating. Diabetics should avoid the use of artichokes due to the high percentage of purine which tends to produce uric acid.

Artichoke, Jerusalem

A root vegetable containing a fairly large amount of inferior (incomplete with only a few amino-acids) protein and sizable amounts of carbohydrates, principally in the form of inulin; also containing high percentage of minerals and an alkali excess; the vitamin content is negligible.

Some doctors will recommend Jerusalem artichokes in place of potatoes due to the inulin not increasing the sugar content in the blood.

Asparagus

Underground sprouts or tips of the plant; very watery, poor in nutritive values; being a bud vegetable, over-rich in acids but containing an alkali excess; rich in an amino-acid known as asparagine; very little vitamins (see chart).

Avocado

A tree fruit. This pulpy fruit is rich in potassium with lesser amounts of sodium and fluorine and iodine; rich in oil and containing large alkali excess. Avocado is rich in vitamin A and has fair amounts of B1, B2, niacin and vitamin C.

Avocados will assist in overcoming constipation and will also help stimulate the appetite.

Bacon

Consisting principally of fat from the sides of the belly of the pig; most of which is cooked out in the process of frying, leaving a residue with a fair amount of acid excess; almost no water-soluble vitamins and only a small quantity of the fat-soluble vitamin A; very little protein which is of the non-assimilable type, thus making bacon possibly the least injurious of all animal food products.

Banana

A plant fruit which is rich in potassium and calcium with a small amount of sodium. The banana has an alkali excess with a trace of vitamin B and a fair amount of vitamin C; very low on fat soluble vitamins. Bananas contain a heavy saturation of starches which, as the fruits ripen, are transformed into sugar and also protein which is incomplete.

Barley

The ripe seeds of the barley grass are rich in potassium, silcic and phosphorus acids and sulphur; showing traces of iodine and fluorine; having high acid excess and almost none of the water-soluble vitamin B which is all concentrated in the seed skins. Barley is very acid forming and thus should only be eaten in small quantities.

Bean, Fresh String

A pod vegetable, very rich in potassium and magnesium; having a rather small alkali excess; containing fair proportions of all vitamins.

Bean, Kidney

A pod vegetable, very rich in protein which is, however, of an inferior incomplete quality; also contains much fat and carbohydrates; rich in potassium, magnesium and phosphorus and sulphur acids; has small amounts of iodine while having an acid excess, only fair amounts of vitamin B and even less of vitamins A and C. Because of their highly concentrated nature, kidney beans should be eaten sparingly.

Bean, Lima

A legume very rich in incomplete protein and carbohydrates; containing per half cup: 63 mg. calcium, 158 mg. phosphorus and 2.20 mg. of iron; very high in vitamin A with good amounts of thiamin, riboflavin and pantothenic acid.

Bean, Soya

A legume containing one of the few non-animal proteins which are complete and have an average of 90.5 percent digestibility. They contain a fairly equal amount of carbohydrates and protein; a very valuable vegetable fat and a high amount of potassium; fair amounts of the vitamins A and B with a trace of vitamin C.

Beef

Steer or heifer meat; rich in calcium and sodium; with high amounts of phosphorus and sulphur acids; having high acid excess; containing small percentage of vitamin B complex, and only a very small amount of vitamin A.

Meat contains a so-called meat alkali, but is composed principally of protein and fat. Beef is considered by many nutritionists as a very good source of protein. Even so, eating beef too often in large quantities is not a healthy habit, as beef contains too much acid-forming protein and so may destroy the necessary alkali excess in the body. For additional information on possible arguments for limited meat eating see (chapter on Vegetarianism).

Beet Greens (Chard)

Leaves of the sugar beet which are very rich in vitamin A and high in potassium and sodium. Like spinach, beet greens contain a considerable amount of oxalic acid which connects rapidly with calcium, forming oxalates which are then excreted.

Beet, Sugar

A root vegetable containing incomplete protein, a plentiful amount of cellulose and natural sugars. High in potassium and sodium; having alkali excess and small amounts of vitamins A, B and C. The sugar beet also contains a large amount of an oxidizing ferment which when it comes into contact with the air changes the tyrosine (amino-acid) found in the juice into a dark insoluble powder.

Blackberry

A bush fruit very rich in sugar and cellulose; high in minerals except sodium; containing high alkali excess; low on vitamin A, traces of vitamin B and only a small amount of C.

Blueberry

A bush fruit containing a small quantity of incomplete protein and a small amount of tannin and pectin; rich in potassium; with a very high alkali excess; no vitamin B and a small amount of vitamin C.

Brazil Nut

Rich in incomplete protein (a high percentage of amino-acids, but not all of the essential ones), oil and fat and starches; also rich in potassium, calcium and phosphorus; high acid excess; with a fair amount of vitamin B and smaller amounts of vitamins A and C.

Brazil nuts should be thoroughly chewed otherwise they tend to irritate the sensitive lining of the intestinal tract.

Bread, Rye

Made from the flour of the complete seeds of rye-grass and Secale cereal (cultivated rye): rich in incomplete protein, oil, starches, potassium, phosphorus and sulphur acids; having a high acid excess and a small amount of vitamin B per slice.

Bread, Wheat

Made from the ground whole grains of the wheat grass; rich in incomplete protein, oil and starches; rich in phosphorus, potassium and sulphur acids; large acid excess and having only a small amount of vitamin B per slice.

The protein content of whole wheat bread is inferior to that of rye bread.

Bread, White

Made from bleached refined flour and grains of the wheat grass with a small amount of incomplete protein and oil; very rich in starches with a little potassium, phosphorus and sulphur acids; possessing acid excess and virtually no natural occurring vitamins. In fact, white bread has very little food value.

Broccoli

A variety of the common cauliflower with the edible parts being the fleshy stalks and pollen; rich in potassium; containing a small amount of phosphorus and sulphur acids and traces of iodine; having an alkali excess; rich in vitamin B and vitamin C.

Brussels Sprout

The bud of the common cabbage plant contains a rich supply of cellulose, potassium, phosphorus acid, and sulphur acid; also rich in incomplete protein; fairly high alkali excess with a fair amount of vitamin A and traces of vitamins B1, B2, B6 and niacin. Brussels sprouts contain sulphur acid and therefore are less valuable as a food than the other varieties of cabbage.

Buckwheat

A grain containing a small quantity of incomplete protein and fat; rich in carbohydrates, potassium, phosphorus and sulphur acids; only traces of the B complex vitamins.

Butter

Unmelted fat extracted from milk; composed of fat, small amounts of calcium, phosphorus and iron; high concentration of vitamins A and D with smaller amounts of thiamin, riboflavin and niacin.

Of all fats, butter is the easiest to digest. Care should be taken to avoid rancid butter which destroys all fat soluble vitamins upon contact. The loss of vitamin A in melted butter is negligible whereas it is completely lost when butter is used in frying.

Cabbage, Red

Has the same composition as white cabbage. The nutritional value of red cabbage is the same as white cabbage.

Cabbage, White

Unopened heads of the white cabbage plant, having a small amount of incomplete protein and little sugar; rich in potassium and calcium; containing large alkali excess and tracess of iodine and iron; rich in water soluble vitamins A and D.

Cabbage, White (Sauerkraut)

White cabbage cut fine and fermented in lactic acid bacilli (brine made from its own juice with salt added); having the same composition as white cabbage with a rich amount of vitamin C. Due to its acid content it has a strong, stimulating effect upon the digestive organs.

Cakes

Made from refined flour with large amounts of sugar and condiments added; having only a small amount of incomplete protein and containing mostly starches; completely deficient in mineral value and having acid excess and virtually no vitamins and therefore no nutritional value.

Carrot

A root vegetable containing a small amount of incomplete protein; rich in carbohydrates, potassium, sodium with a fair amount of calcium and phosphorus, and traces of iodine. Very high vitamin A, only traces of the B-complex and low in vitamin C.

The vitamin A content in every 10-ounce cup of carrots is between 10,000 and 12,000 units. Although young carrots are more tender, larger, more fully mature carrots are not only sweeter but contain a larger amount of carotene; the substance which is converted into vitamin A by enzymes.

Carob

Ripe pod of the Saint-John's bread or carob bean; having a fair amount of incomplete protein and some sugar; rich in potassium, calcium and phosphorus acid with traces of iodine.

There is growing evidence that chocolate (pure cocoa) is actively poisonous; many people are allergic to it while others, exposed from childhood to small increasing quantities build up a resistance. Carob is an excellent replacement for chocolate. The taste buds must be very sophisticated to tell the difference.

Cauliflower

The stalk and flower bud of the brassica family which contains a small amount of incomplete protein; rich in sugar, cellulose, potassium, mustard oil, and calcium; a fair amount of vitamins A, B and C.

Because cauliflower causes mild flatulence it should be steamed or boiled. Also since mustard oil in cauliflower tends to irritate the kidneys, anyone suffering from any form of kidney disease should avoid this vegetable.

Caviar

Eggs of any number of fish, but more commonly from the sturgeon, salmon and haddock; rich in incomplete protein; containing some oil; rich in potassium, calcium, phosphorus and sulphur; high acid excess and moderate amounts of vitamins A and B.

Celery, Bleached

Stalks of the celery plant grown under the surface of the ground; having a slight amount of incomplete protein and a small quantity of natural sugar; good in potassium, fair amounts of sodium and calcium; high alkali excess with only traces of the B-complex, low in niacin and vitamin C.

Celery Leaves

Green leaves of the celery plant growing above the ground; containing a small amount of incomplete protein and a little more sugar than the bleached celery stalk; rich in potassium, sodium and sulphur acid; having a high alkali excess; rich in vitamins A, B and C; also containing an ingredient of insulin and a ferment known as tokokinin.

Chard (See Beet Greens)

Leaves and stalk of the beet plant; containing a fair amount of incomplete protein and carbohydrates; rich in potassium and calcium; having a high alkali excess and rich in the vitamin B-complex.

Cherry

Fruit of a tree, containing a very small amount of incomplete protein; high in natural sugars and fruit acids; very rich in potassium; having an alkali excess; rich in vitamins C and B-complex.

Cherries, eaten daily, tend to regulate the eliminative process and promote a more healthy system. They also stimulate the secretion of urine without injury to the kidneys.

Chestnut

Containing a good source of incomplete protein and fat; very high in carbohydrates, potassium, sodium and magnesium; having an acid excess and only a small amount of the vitamin B-complex.

Cottage Cheese

Curdled and compressed casein (white, crumbling substances with a high acid nature) in milk; very rich in protein, rich in fat, calcium, phosphorus and sulphur acids; having a high acid excess and a small quantity of vitamins A and B-complex.

Cheese, Goat

Made from goat's milk; rich in potassium, protein, fat, phosphorus and sulphur acids; large acid excess and small quantity of vitamin A.

Cheeses

Made from cow's milk; rich in protein, fat, calcium, potassium, phosphorus and sulphur acids; high acid excess and a small amount of vitamin A. Most varieties of cheese are very nutritional.

Chick Peas (Garbanzos)

A legume rich in cellulose, fat and incomplete protein; also having a high concentration of potassium, phosphorus and sulphur acids; an acid excess and small quantity of vitamins A and B-complex.

Chicory

Containing a small amount of incomplete protein; rich in potassium; having an alkali excess and a high content of vitamin A.

Chives

The small, green, spear-like shoots of the chives bulb, containing a rich supply of incomplete protein, oil, potassium and calcium; rich alkali excess and fair quantity of vitamins B-complex and C.

Chives aid in dissolving the phlegm in catarrh and help to stimulate the appetite. It is primarily used as a delicate onion flavoring in salads and soups.

Chocolate

Containing a mixture of sugar and cocoa; rich in incomplete protein, starches and fat; also a fair quantity of potassium, calcium, phosphorus and sulphur acids; having an acid excess with very little vitamins. Chocolate is such a heavily concentrated food that it should only be eaten in small quantities. It also contains an alkaloid, similar to the one found in caffeine, known as theobrama, from which it derives its stimulating properties. Chocolate is a common cause of constipation.

Cider

Fermented apple juice; having only alcohol and fruit sugar; rich in fruit acids and very deficient in minerals; having a slight alkali excess and very little vitamin content.

Apple cider is very beneficial to digestion.

Cocoa

Fermented seeds of the cocoa tree; rich in incomplete protein, starches and fat; very rich in potassium, calcium, phosphorus and sulphur acids; having a high acid excess and a very small quantity of vitamin B.

Cocoa is similar to chocolate in that it too has a stimulating effect and is also fattening. It, also, will cause constipation.

Coffee

Rich in the alkaloid, caffeine, which will cause a stimulating effect; containing virtually no other nutrients.

Strong, black coffee can cause palpitation of the heart and insomnia. It can also cause a minor irritation to the nervous system, when taken in large quantities of three or more cups per day.

Cola

Dried kernels of the cola tree; rich in incomplete protein, carbohydrates, potassium, calcium and phosphorus and sulphur acids; having an acid excess and containing a very small amount of vitamins A and B-complex.

Cola nuts are rich in caffeine and thus have the same stimulating effect upon the system as coffee.

Corn-on-the-Cob

The seeds of this yellow vegetable contain a good supply of incomplete protein, carbohydrates and oil; also rich in potassium and phosphorus and sulphur acids; having an acid excess with almost no vitamins and only slight trace minerals. Foods derived from corn include corn-oil and corn flour.

White corn is inferior to yellow corn and is also not easily digestible. Corn is the least nutritious of all grains.

Crab Meat

The meat of a shellfish, containing a good source of complete protein, with very little fat; rich in potassium, sodium and phosphorus and sulphur acids; also having a high amount of iodine and a trace of vitamins B1 and B2; having a high acid excess.

Crackers

Made with wheat flour and salt. The composition and food value of crackers are the same as those of wheat bread.

Cranberry

A berry containing very little incomplete protein but fairly rich in fruit acids, especially in tannic acid; also a fair amount of potassium, and a small amount of calcium, phosphorus and sodium; traces of vitamins C and B-complex; having an acid excess.

Cream

Fat, yellowish part of milk, which rises and collects on the surface of milk; containing a small amount of complete protein; a good amount of milk-sugar; rich in fat, potassium, calcium and phosphorus and sulphur acids, having a small acid excess; rich in vitamin A with lesser quantity of vitamins B-complex and D. Cream is a very nutritious food.

Cucumber

A vegetable of the squash family, having a small incomplete protein content and a fair quantity of sugar and cellulose; rich in sodium, calcium and magnesium; very rich in phosphorus and sulphur acids; having an alkali excess; small amount of vitamin C, with a lesser amount of vitamin B-complex.

Cucumbers are an excellent food except when they have been salted or pickled, in which case they lose most of their food value.

Currant

A berry, containing small amount of incomplete protein; rich in sugar and potassium; having an alkali excess and a fair amount of vitamin B-complex with a lesser amount of vitamin C. Currant juice is especially beneficial in cases of stomach or intestinal catarrh.

Dandelion Green

Young leaves from the common weed, containing a small quantity of incomplete protein and also having a small amount of tannic acid and bitter ingredients (natural mineral salts and juices, which cause a bitter taste); very rich in potassium, calcium, sodium and phosphorus acid; having a high alkali excess and fairly rich in vitamins B-complex, C and D and an extremely high amount of vitamin A. Dandelion greens stimulate the glands and cause a copious flow of bile and act as a strong urine dissolvent.

Date

Fruit of the date palm, containing a fair amount of incomplete protein; very rich in sugar and cellulose; also rich in sodium, magnesium, potassium and calcium; rich in sulphur and chlorine; having an alkali excess and a small amount of vitamin B-complex.

Dill

An herb containing a fairly rich supply of incomplete protein, carbohydrates and cellulose; very rich in ethereal oils, potassium, sodium, calcium and phosphorus and sulphur acids and chlorine; having a strong alkali excess; rich in vitamins A, B, and C. Owing to the heavy saturation of ethereal oil, dill should not be eaten by those who have kidney disorders.

Eggplant

A bulbous, purple-skinned vegetable of the squash family containing a small quantity of incomplete protein and carbohydrates; having a fair amount of minerals and an alkali excess; rich in vitamins A, B-complex and C.

Eggs

Very rich in complete protein and fat; rich in potassium, sodium, chlorine, and phosphorus acid; having an acid excess and a small amount of vitamins A, B-complex and D.

Eggs are overestimated as a food, probably due to their high protein content by weight. There has also been a misconception about the ease of digestion of the soft-boiled egg. In reality, hard-boiled eggs are much less difficult to digest. Eggs contain approximately 6 usable protein grams per 2 ounce (or 1 egg).

Egg White

Rich in complete protein; also rich in potassium, sodium and sulphur acid and chlorine; having an acid excess and very little vitamin B-complex.

Egg Yolk

Containing a rich amount of complete protein, fat and lecithin; rich in potassium, sodium, calcium and sulphur; containing a very large amount of phosphorus; having an acid excess and a small amount of vitamins A, B-complex and D.

Endive

A vegetable plant used in salads which contains a small quantity of incomplete protein and carbohydrates; rich in potassium and calcium; having an alkali excess and a rich supply of vitamin B-complex.

Fennel

A plant with aromatic seeds used as flavoring, containing a small quantity of incomplete protein and a small amount of carbohydrates; having an alkali excess with a rich supply of minerals. Fennel is principally used as a condiment, or as a flavoring for foods. It is only used in very small quantities.

Fig

A tree fruit, containing a poor supply of incomplete protein but very rich in natural fruit sugar, calcium, potassium, magnesium and phosphorus acid; having an alkali excess and a rich supply of vitamin B-complex. Figs contain a beneficial protein-decomposing ferment called Papain.

Figs also stimulate the digestive organs and are one of the best natural purgatives.

Fish

Flesh of fish is far easier to digest than any animal meat; it is also very rich in complete protein; little oil (although there are some oil-heavy fish) and hardly any carbohydrates; rich in potassium, sodium, iodine, and phosphorus and sulphur acids; having an acid excess.

Fowl

Poultry meat has basically the same make-up as animal meat; rich in potassium and sulphur and phosphorus acids; having an acid excess.

Garlic

A bulb, the root of the garlic plant, containing a small amount of incomplete protein; rich in sugar and raw fiber; fair amount of mustard oil; very rich in potassium, calcium and phosphorus acid; having an alkali excess and traces of iodine; fairly rich in vitamins A, B-complex and C.

Garlic stimulates the appetite and acid secretion of the gastric juices and thus helps prevent gas formation or flatulence. Garlic also promotes peristaltic action or the movement of the bowels and so tends to heal inflammatory conditions of the intestines, created by constipation.

Ginger

Roots of the ginger plant, containing a small amount of incomplete protein, carbohydrates and ethereal oil; rich in mineral elements; having an alkali excess; it is used as an aromatic flavoring.

Ginger increases the appetite and aids digestion, helps to regulate delayed menstruation and has a fairly strong diuretic action.

Grapefruit

A hybrid citrus fruit, containing a poor supply of incomplete protein and carbohydrates; very rich in fruit acids and their salts; especially high in citric acids; rich in potassium, vitamin B-complex and C and having an alkali excess. Grapefruit serves as a natural stomach bitter which increases the flow of digestive juices.

Grape

A vine fruit, containing a poor amount of incomplete protein; rich in fruit acid salts and vinous acid; very rich in sugar and potassium; having an alkali excess and a rich supply of vitamins B-complex and C.

Grapes increase and normalize urine and lessen the formation of uric acid while increasing the excretion of uric acid from the body and decreasing the amount of acid in the body. Grapes eaten regularly also assure better intestinal elimination.

Ham

Smoked or otherwise cured pork, containing a rich source of complete protein and fat; also rich in sodium, potassium and phosphorus and sulphur acids; having an acid excess and a fair amount of B1, B2 and niacin.

Due to its method of curing with kitchen salt and by smoking, ham is more injurious than any other kind of meat. It should be eaten no more than one meal per week and only in small quantity.

Hazelnut (Also known as Filbert)

A nut, containing a rich supply of complete protein; small quantity of carbohydrates; rich in potassium, calcium, phosphorus and sulphur acids; having an acid excess and a fair amount of vitamins A, B-complex and C.

Honey

Nectar of flowers transformed into honey through digestion in the ante-stomachs of honey bees; containing a rich supply of grape and fruit sugar with a trace of formic acid; poor in minerals; having a slight acid excess and traces of vitamins B-complex, C, D, and E; containing various animal ferments, especially oxydase (oxydizing ferment).

Honey causes a slight purgative action. It also possesses a diuretic effect which does not harm the kidneys.

Juniper Berry

A bush berry, containing a rich supply of incomplete protein, sugar, tannin, ethereal oils, potassium, magnesium and phosphorus acid; having an alkali excess and rich in vitamins B-complex and C. It is also popular as a condiment and has a slight diuretic action.

Kale

Leaves of the open cabbage which contain a small amount of incomplete protein and some carbohydrates; a good supply of potassium and calcium, some iron, phosphorus and sodium; having an alkali excess, a very high amount of vitamin A, and a rich but lesser amount of vitamins B-complex and C.

Kale, Sea

Sprout and young leaf of the sea kale; containing a fair amount of complete protein and a lesser amount of sugar; rich in potassium, sodium and iodine; having an alkali excess with a very good supply of vitamin A and a fair amount of vitamins B-complex and C.

Lamb

Meat of young sheep; rich in complete protein, fat and gluten; also a good supply of potassium, phosphorus, less of sodium, calcium and iron; and a fair amount of sulphur acids; having an acid excess with some vitamin B-complex.

Lard

Melted fat extracted from the waste parts of the pig; having few minerals and an acid excess and virtually no vitamins. Vegetable oils are far superior and much more nutritious than lard.

Leek

A bulb and green shoot having onion-like flavor and containing a small amount of incomplete protein, some carbohydrates and a slight trace of mustard oil; rich in potassium, sodium and phosphorus acid; having an alkali excess and rich in vitamins B-complex, C and only a small amount of A.

Leeks are beneficial in that they promote urine secretion and help prevent flatulence. They also help inflammation of the air passages due to colds because they loosen phlegm. They also increase the appetite and stimulate the digestive glands.

Lemons

A citrus fruit, containing small amounts of incomplete protein; very rich in citric acid and acid salts; rich in sugar, potassium and calcium; having an alkali excess and fairly rich in vitamin C with only a trace of B2. Lemons have a diuretic action without injury to the kidneys. They are recommended for mouth sores and swollen gums.

Lentil

A legume, containing a rich supply of incomplete protein and carbohydrates; also rich in cellulose, purine, potassium, sodium and calcium; over-rich in phosphorus and sulphur acids; having an acid excess and traces of vitamins A and B-complex.

Lettuce, Head

A salad vegetable containing some caoutchouc (rubber), chyle and bitter tasting ingredients; rich in potassium, calcium, sodium, iron with traces of iodine; having an alkali excess and fair amount of vitamin B-complex, traces of vitamin C and a good amount of vitamin A. The darker green, outer leaves contain the most vital food elements. Many types of lettuce are available including iceberg, Boston, Bib and Romaine. Only one is outstanding in its high content of vitamin A and that is Romaine.

Lime

A citrus fruit, containing a small quantity of incomplete protein and sugar; very rich in citric acid and citric acid salts; fair amount of potassium and calcium; having an alkali excess and a sizable quantity of vitamin C and a fair to low amount of vitamin A.

Liver

Containing a very rich source of complete protein, potassium, sodium and magnesium, also rich in phosphorus and sulphur acids; having an acid excess and a high amount of the B-complex and a very rich source of vitamin A; also a moderate amount of vitamin D. Beef liver has about a third more vitamin A than either calf or chicken livers. (¼ lb. Beef 46,000 I.U. of A; ¼ lb. Calf 32,000 I.U. of A; ¼ lb. Chicken 32,200 I.U. of vitamin A.)

Lobster

A shellfish rich in complete protein; with a small amount of fat; low in sodium, fair in potassium and phosphorus and sulphur acids; also contains a slight trace of iodine; having an acid excess and a small quantity of vitamin incomplete B-complex.

Macaroni

Pasta, a form of noodles made from the meal or grits of wheat flour mixed with water and eggs to make a dough; containing a fair quantity of incomplete protein, small proportion of fat and a rich source of starches; also rich in potassium and phosphorus and sulphur acids; having an acid excess and a small amount of vitamin B-complex.

Malt

Germinated seeds of different grains, but usually barley; containing an incomplete protein and small quantity of fat and starches; rich in malt-sugar, potassium and phosphorus and sulphur acids; having an acid excess and fair amount of vitamin B-complex.

Mango

A pulpy, tropical fruit, containing a very small amount of incomplete protein and a high sugar and cellulose content; having an alkali excess and fair amounts of vitamin A and small amounts of vitamin B1, B2 and C.

Maple Syrup

Syrup boiled down from sap of the maple sugar tree; rich in potassium, sugar and calcium; having an alkali excess and rich in vitamin B-complex.

Margarine

Fat of the same consistency as butter, obtained by pressing melted vegetable fats, which have been hydrogenated and made to look like butter. Containing some mineral elements; having an acid excess and a fair amount of vitamin A, due to fortification with a synthetic. Hydrogenating the oils to make them solid destroys most of the essential fatty acids.

Melon

A pulpy fruit, of which there are several varieties, related to squash and pumpkin, containing a small amount of incomplete protein; rich in sugar, cellulose, potassium and having an alkali excess; low in vitamins B-complex and C. Melons also have a mild diuretic action which does not harm the kidneys.

Milk (from Cows)

Containing a small amount of alkali excess; very rich in complete protein, fat and sugar; low in potassium and sodium, very rich in phosphorus and sulphur acids; also rich in vitamins A, B-complex, fair C and D. Milk possesses oxidizing ferments which are not always easy to digest. For babies and small children raw cow's milk sometimes contains too much protein and should be diluted with water. Also in the boiling or sterilizing of milk, various changes occur which affect the qualities of the milk. The vitamin content is changed, the valuable protein is partially decomposed and the bone-building salts in colloidal-loose form are changed to a combined form which is difficult to assimilate.

Milk, Butter

Slightly curdled milk from which the fat has been removed. Its composition is almost the same as sour milk with the exception that it is deficient in fat.

Milk, Skimmed

Milk from which the fat has been skimmed or separated. Composition is much the same as regular milk except that it contains only traces of fat.

Milk, Sour

Milk curdled due to the formation of lactic acid; the casein being transformed into gelatinous masses. The composition of sour milk is the same as ordinary, sweet milk, except that it has less sugar.

Sour milk has been used in cases of stomach or intestinal ailments.

Millet

A grain, containing large quantities of incomplete protein; some fat and rich in starches, potassium, magnesium and phosphorus and sulphur acids; having an acid excess and small amount of vitamin B-complex.

Molasses

Uncrystallizable, evaporated syrup obtained from the raw cane sugar residue when reduced to a fluid by boiling; rich in raffinose, sucrose, grape and fruit sugars; having no protein and very rich supply of minerals; having an alkali excess and only a trace of vitamin B1.

Mushroom (Any of a Variety of Fleshy Fungi)

Containing a rich supply of incomplete protein; some carbohydrates; rich in sodium, potassium, and sulphur; having an alkali excess and a little Vitamin A and D with traces of the B-complex.

Muskmelon

A fruit, containing a small quantity of incomplete protein and a fair amount of sugar; rich in potassium; having an alkali excess and a fair supply of vitamin B-complex; also possessing a strong diuretic action without injuring the kidneys.

Mustard Greens

Leaves of the mustard plant, containing an incomplete protein and small amount of carbohydrates and a rich supply of minerals; having an alkali excess; high in vitamin A, small amount of B-complex, and a fair C content.

Mutton

Flesh of the fully grown sheep; rich in complete protein and fat; very rich in potassium, and phosphorus and high in sulphur acids; having an acid excess with a small amount of vitamins A and B-complex. Mutton fat is very difficult to digest.

Nectarine

A tree fruit, containing a rich amount of carbohydrates and sugar; a small quantity of minerals; having an alkali excess and a small amount of vitamin C.

Nuts

Virtually every nut is rich in protein, fat and carbohydrates; also in potassium, sodium, calcium and phosphorus and sulphur acids; having an acid excess. (See chapter on Nuts & Dried Fruit.)

Oats

A grain, containing a rich supply of incomplete protein, starches, potassium, phosphorus and sulphur and silicious acids; having an acid excess and small quantity of vitamins A and B-complex.

Okra

A vegetable, a tall plant with mucilaginous green pods, containing a rich supply of incomplete protein and carbohydrates; also rich in sodium and calcium; having an alkali excess and small amount of vitamin B-complex. Okra has been used for aiding digestion and helping to overcome inflammation of the stomach and intestines.

Olive

A tree fruit, having a large quantity of incomplete protein and carbohydrates; very rich in potassium, fat, sodium, and calcium; having an alkali excess and small amount of vitamins A and B1. The cold pressed oil of olives is very healthful and contains more nutritional properties than any other vegetable oil. Olives also contain more protein than any other fruit.

Onion

A root-bulb vegetable, containing a small quantity of incomplete protein and a fair amount of sugar; very rich in ethereal mustard oil, potassium and calcium; having a very rich supply of sulphur and an alkali excess; possessing a fair amount of vitamins B-complex and C. Onions are strongly diuretic in their action.

Orange

A citrus fruit, containing a small amount of incomplete protein and a fair supply of sugar; very rich in citric acids and fruit acid salts; rich in calcium and potassium; having an alkali excess and a rich source of vitamins C, B-complex and A.

Oyster

A shellfish, containing a rich source of complete protein and having little fat; rich in sodium, chlorine and potassium; also rich in phosphorus and sulphur acids; strong traces of iodine; having an acid excess and a good supply of vitamins A and B-complex.

Papaya

A pulpy fruit, fairly rich in incomplete protein, sugar and fat; very rich in sodium and magnesium; fairly rich in phosphorus and sulphur; having an alkali excess; a good source of papain, which aids digestion. High in vitamin A, with traces of niacin and good C.

Paprika

A mild, powdered seasoning made from sweet red peppers; contains virtually no nutritional properties and can cause an irritation to the mucous membranes and the intestinal tract.

Parsley

An herb and a vegetable, containing a trace of incomplete protein and ethereal oil; rich in calcium, potassium and magnesium; having an alkali excess and a large quantity of vitamins B-complex and C and a very high amount of vitamin A. Parsley is diuretic in its action. It has also been used in cases of menstrual irregularities. Stimulates digestion.

Parsnip

A root vegetable containing a small quantity of incomplete protein; fairly rich in sugar and very rich in ethereal oil, cellulose, potassium and calcium; having an alkali excess and small amounts of vitamins A and B-complex.

Peach

A tree fruit, having a fair amount of incomplete protein and sugar; very rich in calcium and potassium; having an alkali excess and fair quantities of vitamins B-complex and C.

Peaches aid digestion; they are also diuretic and laxative.

Peanut

A ground-nut, containing a very rich supply of incomplete protein and fat; fair amounts of carbohydrates and purine; very rich in potassium, calcium, magnesium and phosphorus and sulphur acids; having an acid excess and a fair amount of vitamins A and B-complex.

Pear

A tree fruit, having a small quantity of incomplete protein; rich in sugar, and potassium; having an alkali excess and a small amount of vitamins B-complex and C.

Pea, Fresh

A legume vegetable, containing a fairly rich supply of incomplete protein, sugar and starches; very rich in magnesium, potassium, phosphorus and sulphur acids; having an alkali excess and a small amount of vitamins A, B-complex and C.

Pecan

A tree nut, containing the highest oil content of all nuts. They have about 71 percent oil. They also contain all of the elements mentioned under the *Nut* listing. They have a mild laxative effect and should be chewed well before swallowing.

Pepper, Black

Spice, ground from seed, used for flavoring and containing a rich supply of acrid ingredients which cause an irritation to the mucous membranes, the kidneys and urethra.

Pepper, White

Spice, ground from seed, used for flavoring and containing about twice as much acrid-tasting ingredients as the black pepper, but white pepper is milder because it contains less acrid resins.

Persimmon

A fruit, having a small quantity of incomplete protein and a good supply of carbohydrates; rich in potassium and fair in phosphorus acid; having an alkali excess, high vitamin A and some C. Persimmons have a strong purgative action.

Pickles

Small cucumbers or other vegetables or fruits, preserved in vinegar or mustard. They have no beneficial food value and thus should not be eaten regularly.

Pineapple

A bush fruit, rich in potassium, malic, citric and tartaric acids; having a high percentage of cellulose and a ferment known as papain, which neutralizes protein acids; having an alkali excess.

Fresh pineapples have been used for poor digestion and constipation. They supply the weak stomach with the lacking acid salts and act as a disinfectant for the digestion of food. They have a diuretic action.

Pine Nut (Pignolia)

Containing the highest percentage of protein of any nut (33.9 percent) (see chapter on *Nuts*); also rich in mineral elements, fat, and tannic acid; having an acid excess and small amount of vitamin B-complex.

Plum

A tree fruit, containing a fair amount of incomplete protein; very rich in sugar and some fruit acid salts; high content of potassium; having an alkali excess and a small amount of vitamin B-complex. Plums are beneficial for relief of constipation and hemorrhoids. Due to their stimulating effect upon the intestines they have also been used in liver ailments.

Pork

Flesh of the domestic pig; rich in complete protein and fat; very rich in sulphur acids and phosphorus; having an acid excess and only traces of vitamins A and B-complex.

Potato

A root vegetable, containing a small amount of valuable protein and rich in starches; very rich in potassium and traces of iodine; having an alkali excess and small amounts of vitamins A, B-complex and C.

Potato, Sweet

A root vegetable, containing a small quantity of incomplete protein with larger amounts of sugar and starches; very rich in calcium and potassium with traces of iodine; having an alkali excess and containing a very good supply of vitamin A, with a small amount of vitamins B-complex and C.

Pumpkin

A vegetable of the squash family, containing a poor amount of incomplete protein; fairly large quantity of sugar and cellulose; very low in sodium and a fair amount of potassium; having an alkali excess, excellent vitamin A and an adequate supply of B-complex and C.

Quince

A fruit, having a low supply of incomplete protein, high sugar and fruit acid salts, also some tannic ingredients; fair in cellulose, potassium and pectin; having an alkali excess; low vitamin A, B-complex and C.

Radish

A root vegetable, having a poor supply of incomplete protein; fairly rich in sugar and mustard oil; having a large quantity of cellulose and potassium; having an alkali excess and small quantities of vitamins B1, B6 and C.

Radishes are strongly diuretic. They also stimulate the appetite and digestion.

Radish, Horse

A root vegetable, containing a small amount of incomplete protein and only traces of fat; having a large quantity of carbohydrates and raw fiber and ethereal mustard oil; very rich in calcium, potassium and phosphorus acid; also rich in sulphur acid; having an alkali excess and no known vitamins.

Mustard oil tends to irritate the kidneys, the bladder and mucous membranes of the digestive tract. Horse radish does stimulate the appetite and digestion and aids in loosening phlegm.

Raisin

The dried fruit of the grape, containing a rich supply of incomplete protein, sugar and cellulose; also rich in calcium, potassium, magnesium and phosphorus and sulphur acids and chlorine with traces of iodine; having an alkali excess and very small amount of vitamin B-complex.

Raspberry

A bush berry, having a small amount of incomplete protein and a large supply of sugar; rich in vitamins C and B-complex with an alkali excess.

Rhubarb

A vegetable (but eaten cooked as fruit), containing a small amount of incomplete protein, sugar and fruit acid salts, particularly oxalic acid; good in potassium; having an alkali excess, low in vitamin A, and only a trace of vitamins C and B-complex.

Romaine

Has the same composition as that of common lettuce with the exception of almost double the amount of vitamin A.

Spinach

A leafy vegetable, containing a good supply of incomplete protein; rich in mineral salts and having an alkali excess. Raw spinach juice has a fair amount of iron, excellent vitamin A, low B-complex and C.

Squash

A pulpy vegetable, containing a small amount of incomplete protein; rich in minerals, particularly potassium; having an alkali excess and a rich source of vitamin B-complex, a good supply of A and a small amount of C.

Strawberries

A berry, containing a rich fruit sugar and potent bactericidal elements; having an alkali excess with some minerals, incomplete protein and a fair supply of vitamin C.

Sugar

Containing virtually no food value and the most overrated mass-consumed food item in America. All sugars belong to the carbohydrate group. The unprocessed, unrefined, raw sugar is better than the white sugar which is too concentrated and acid-forming. Sugar should not be used. Raw honey makes an excellent substitute for sugar and has none of the harmful elements of refined sugar.

Tomatoes

A berry (but eaten as a vegetable), containing a rich supply of potassium, low in other minerals; also rich in vitamin A, good in B-complex and C; and having an alkali excess.

Turnips

A root vegetable, having much the same composition as potatoes and can be eaten in place of them.

Vinegar

Wine vinegar made from apple cider; having an acid excess and no beneficial food elements; lemon juice should be used in place of vinegar.

Walnut

Containing a rich and incomplete protein, fat and starches; small amount of calcium, fair amount of potassium and magnesium and phosphorus and sulphur acids; having an acid excess and small amount of vitamins A and B-complex.

Yogurt

A custardlike food prepared from milk curdled by bacteria and often sweetened or flavored with fruit; containing a good supply of complete protein; rich in phosphorus and sulphur; low in potassium and sodium; very low in fat (about 96 percent fat free); because of its oxidizing ferments and natural beneficial bacteria, yogurt is easy to digest; rich in vitamin B1, traces of D and C.

INDEX